NUTRITION AND GENOMICS

Issues of Ethics, Law, Regulation and Communication

NUTRITION AND GENOMICS

Issues of Ethics, Law, Regulation and Communication

Edited by

DAVID CASTLE

Institute for Science, Society and Policy, University of Ottawa, Canada

NOLA M. RIES

Health Law Institute, University of Alberta, Canada

AMSTERDAM • BOSTON • HEIDELBERG • LONDON • NEW YORK • OXFORD
PARIS • SAN DIEGO • SAN FRANCISCO • SINGAPORE • SYDNEY • TOKYO

Academic Press is an imprint of Elsevier

Academic Press is an imprint of Elsevier
30 Corporate Drive, Suite 400, Burlington, MA 01803, USA
32 Jamestown Road, London NW1 7BY, UK
525 B Street, Suite 1900, San Diego, CA 92101-4495, USA
360 Park Avenue South, New York, NY 10010-1710, USA

First published 2009

Library of Congress Cataloging in Publication Data
A catalog record for this book is available from the Library of Congress

British Library Cataloguing in Publication Data
A catalogue record for this book is available from the British Library

ISBN: 978-0-12-374125-7

For information on all Academic Press publications visit our website at www.elsevierdirect.com

Printed and bound in the United States of America

09 10 11 12 13 10 9 8 7 6 5 4 3 2 1

Contents

List of Contributors

Laura Bouwman Communication Science (bode 79) Wageningen University, PO Box 8130, 6700 EW, Wageningen, the Netherlands

Tania Bubela School of Public Health, University of Alberta, 13 – 106 D Clinical Sciences Building, Edmonton AB T6G 2R6, Canada

Alan Cassels 423 Standard Avenue, Victoria, British Columbia V85 3M6, Canada

David Castle Department of Philosophy, University of Ottawa, Ottawa, Ontario K1N 6N5, Canada

Timothy Caulfield Faculty of Law, Health Law Institute, Law Centre, University of Alberta, Edmonton ABT6G 2H5, Canada

Ruth DeBusk PO Box 180279, Tallahassee, FL 32318-0279, USA

Jennifer Farrell 229 SMD, University of Ottawa, Ottawa, Ontario K1N 6N5, Canada

Rosalynn Gill Sciona, Inc., 12635 E. Montview Blvd, STE 217, Aurora, CO 80045, USA

Peter J. Gillies DuPont Haskell Laboratory for Health and Environmental Sciences, 1090 Elkton Road, Newark, DE 19714-0050, USA

Stuart Hogarth University of Cambridge, Department of Public Health Primary Care, Cambridge, UK

Michiel Korthals Philosophy, Wageningen University, Hollandseweg 1, 6706 KN, Wageningen, the Netherlands

Elaines S. Krul Solae LLC, 4300 Duncan Ave., St Louis, MO 63110-1110, USA

Leia Minaker Centre for Health Promotion Studies, University of Alberta, 5-10 University Terrace, 8303-112 Street Edmonton, AB T6G 2T4, Canada

Karine Morin Department of Philosophy, University of Ottawa, Ottawa, Ontario K1N 6N5, Canada

Jose M. Ordovas Tufts University, HNRC, 711 Washington Street, Boston, MA 02111-1524, USA

Nola M. Ries Research Associate, Health Law Institute, University of Alberta & Adjunct. Professor, Faculty of Human and Social Development, University of Victoria, Sedgewick B122, PO Box 1700, STN CSC, Victoria, British Columbia V8W 2Y2, Canada

Milly Ryan-Harshman FEAST Enterprises, Nutrition Consultant & Owner; Sessional Instructor, Faculty of Health Sciences and Faculty of Science, University of Ontario, Institute of Technology, 2000 Simcoe Street, North Oshawa, Ontario, Canada, L1H 7K4

Paula Saukko Department of Social Sciences, Brockington Bldg., Loughborough University, Leicestershire, LE11 3TU, UK

Jacob Shelley BA, MTS, LLB, LLM (C) Researcher, Health Law Institute, University of Alberta, Edmonton ABT6G 2H5, Canada

E. Shyong Tai Department of Endocrinology, Singapore General Hospital, Department of Medicine and Center for Molecular Epidemiology, National University of Singapore, Singapore

Cees Van Woerkum Communication Science (bode 79) Wageningen University, PO Box 8130, 6700 EW, Wageningen, the Netherlands

Ellen Vogel Faculty of Health Sciences, University of Ontario Institute of Technology, 2000 Simcoe Street, North Oshawa, Ontario, L1H 7K4, Canada

Acknowledgments

We have several important acknowledgements to those who made this book possible. First, we thank the authors who agreed to contribute to this book and dedicated their time to researching, analyzing and articulating current issues in nutrigenomics in areas of science, ethics, law, regulation and communication. We asked contributors not only to identify those key issues, but also to present options for addressing them and make recommendations for future steps relevant to scientists, industry, regulators, health care professionals, the media and other stakeholders in nutrigenomics. As part of this project, contributors participated in a workshop in Banff, Canada, in January 2008 and we thank them, especially international contributors, for traveling to be part of this important meeting where we discussed and debated the book's content.

Second, we acknowledge generous funding support from the Advanced Foods and Materials Network (AFMNet), one of Canada's Network of Centres of Excellence. AFMNet, a network of scientific researchers, professionals, industry partners and government agencies, provided funding for a 3-year (2006–2009) project examining ethical, legal and social issues in nutrigenomics. This project, titled 'Social Issues in Nutritional Genomics: The Design of Appropriate Regulatory Systems and Issues of Public Representations and Understanding', is co-led by David Castle, Canada Research Chair in Science and Society, University of Ottawa, and Timothy Caulfield, Canada Research Chair in Health Law and Policy, University of Alberta. This book is a key outcome of the project, along with a variety of other publications and knowledge translation activities.

Finally, we express great thanks to Susan Callan, Program Manager, Institute for Science, Society and Policy, University of Ottawa, for her work in administering the project and ensuring that meetings, travel arrangements and other administrative matters went smoothly. We also thank the students and trainees who assisted many contributors to this book. In particular, we are grateful for research assistance from Juliana Aiken and Sarah Scott, University of Ottawa, and Victor Alfonso, Christopher Ali, Maria Chau, Laura Geddes, Simrat Harry and Ben Taylor, University of Alberta.

David Castle and Nola M. Ries, Co-Editors

Editors' Introduction

In 1975, Irma S. Rombauer and Marion Rombauer Becker, authors of the iconic cookbook, *The Joy of Cooking,* wrote:

> How we wish someone could present us with hard and fast rules as to how and in what exact quantities to assemble the proteins, fats and carbohydrates as well as the small but no less important enzyme and hormone systems, the vitamins, and the trace minerals these basic foods contain so as to best build body structure, maintain it, and give us an energetic zest for living!
>
> [...]
>
> Individually computerized diagnoses of our lacks may prove a help in adjusting our deficiencies to our needs. But what we all have in our bodies is one of the greatest of marvels: an already computerized but infinitely more complex built-in system that balances and allocates with infallible and almost instant decision what we ingest, sending each substance on its proper course to make the most of what we give it.

Written nearly 35 years ago, these words foreshadow nutrigenomics, the rapidly developing field of science that investigates the interaction between the human genome and nutrients in our diets. *The Joy of Cooking* authors also noted the limitations of public nutrition recommendations for daily intake of nutrients: '... no one chart or group of charts is the definitive answer for most of us ... Such [nutrition] studies are built up as averages, and thus have greater value in presenting an overall picture than in solving our individual nutrition problems.' Nutrigenomics, however, aims to give more precise nutritional advice to groups or individuals in light of their genetics.

Nutrients comprise the greatest and most continuous source of environmental influence on our genomes. Nutrient–gene interactions are involved in regulation of critical pathways that keep us alive and healthy; other influences are deleterious and are implicated in major chronic diseases. Understanding which nutrients have these beneficial or harmful effects for specific genotypes is first among the three core objectives of nutrigenomics. The second objective of nutrigenomics is to understand the implications of nutrigenomics for the development and progression of chronic diseases, such as heart disease, diabetes, arthritis and cancer.

With this knowledge, the third objective emerges: the ability to prevent the onset, or mitigate the effects of, chronic disease. This objective is based on the complicated dynamics of two factors, genetic variation and differences in nutrient exposure. These factors interact in real life, sometimes even with instruction from *The Joy of Cooking,* not in the controlled environment of the laboratory.

The innovation of new science and technology is often thought of in metaphorical terms as a life cycle. This is construed in different ways, sometimes as the flow from research to development and then delivery, or as a product innovation cycle in which innovative products displace conventional products in the marketplace. Similarly, regulators and policy makers often think in terms of life cycles to reflect the different stages through which science and technology interact with government agencies and require foresight, public consultation, technology assessment and monitoring and surveillance.

In selecting and organizing content for this book, we adopt a life cycle view of innovation in nutrigenomics. We recognize that in addressing the different aspects of the life cycle of nutrigenomics, we face two challenges. First, nutrigenomics has already undergone its first full cycle from the basic concepts underlying nutrient–gene interactions through to the first commercial offerings and monitoring by regulators. Nutrigenomics is a scientific, commercial and regulatory reality. The second challenge is that nutrigenomics is a moving target. The science and technology that underpin it change rapidly, the commercial marketplace is in constant flux and regulators and policy makers struggle to keep up with these changes. At the same time, these challenges are precisely what make nutrigenomics an interesting field in which to study issues of ethics, law, regulation and communication.

This book begins with a discussion of nutrigenomics science and technology in Ordovas and Tai's chapter, *Gene–environment interactions: where are we and where should we be going?* Using the example of nutrigenomics, lipids and cardiovascular disease, Ordovas and Tai discuss the most promising research on nutrient–gene interactions. They provide insight into why these nutrient–gene interactions are important, but also point out some of the deficiencies in the science, for example the size and power of studies and the basic techniques used to collect dietary intake data. Many of these issues will need to be resolved before this science can support public applications. In the second chapter, *Translating nutrigenomics research into practice: the example of soy protein,* Krul and Gillies discuss how complicated the translation of research into practice can be. The example of soy protein reveals that many of the important bioactive proteins in soy are poorly characterized in nutrition studies and so claims about their effects are difficult to repeat. Doubts about the quality of nutrition studies will not disappear with the overlay of genetic factors, leading Krul and Gillies to conclude that there needs to be careful control over the quality of the science to improve the 'signal to noise' ratio.

There is a nutrigenomics 'industry' in the sense that commercial applications of nutrigenomics have existed for several years. The chief scientific officer of Sciona, one of the first companies to deliver nutrigenetic testing to the public, offers her views in the third chapter, *Business applications of nutrigenomics: an industry perspective.* Gill provides an insider's view of the business opportunities and constraints facing the nutrigenomics entrepreneur. Without question, one of the most important concerns for the innovator is the regulatory environment, especially regulatory uncertainty that often characterizes new areas of science and technology. Hogarth discusses one significant aspect of this in his chapter, *Regulation of genetic tests: an international comparison,* noting that governments in different countries are weighing options for enhancing their regulation of genetic tests in light of concerns raised about the proliferation of, and greater access to, genetic tests.

The lightning rod is direct-to-consumer genetic tests, offered by Sciona and other firms. These attract much regulatory debate, as Ries discusses in the chapter, *Risk-based regulation of direct-to-consumer nutrigenetic tests.* The manner in which nutrigenetic products and services are regulated – or left largely unregulated – will significantly affect the business strategy of nutrigenomic companies. Regulatory issues in nutrigenomics raise broader questions about legislative frameworks governing food, drugs and medical devices. Morin's chapter, *The impact of genomics on innovation in foods and drugs: can Canadian law step up to the challenge?*, examines the way in which genomics will drive changes in Canadian food and drug law and considers whether basic legislation will have to be reformed to handle innovations in science and technology.

Translating nutrigenomics knowledge into practice to achieve personal and public health benefits requires analysis of factors that influence dietary and other lifestyle behaviors. In the chapter, *Placing healthy eating in the everyday context: towards an action approach of gene-based personalized nutrition advice,* Bouwman and van Woerkum reflect on how nutrigenomics can be incorporated into dietary advice and raise questions about whether it will lead to improved dietary practices. As it stands, it is unknown whether the genetic aspect of personalized nutrition will improve outcomes. This same question faces health care practitioners, particularly dietitians, who are likely to have clients and patients seeking information about nutrigenomics. As nutrigenomic applications come available, appropriate training for health care professionals will be a critical component for translating research to the clinic and the public. Farrell's chapter, *Health care provider capacity in nutrition and genetics* – *a Canadian case study,* explores capacity within Canadian health professions training for incorporating nutrigenomics into professional education and practice. The prognosis is not good since most health care practitioners receive little training in nutritional sciences or genetics, and none in nutrigenomics. Vogel, DeBusk and Ryan-Harshman, recognizing this problem, consider what would be required to train dietitians. Their chapter, *Business advancing knowledge translation in nutritional genomics by addressing knowledge, skills and confidence gaps of registered dietitians,* identifies five gaps in dietitians' knowledge to be addressed in curriculum changes if these health care professionals are going to incorporate nutrigenomics into their professional practice.

As an emerging field, nutrigenomics raises challenges in communicating complex information to the public about the relationship between genomics, genetics and nutrition. This challenge is influenced by growing public reliance on the Internet for health information and decision-making independent of primary health care counseling. These issues are addressed in a trio of chapters, beginning with Cassels' chapter, *Understanding hopes and concerns about nutrigenomics: Canadian public opinion research involving health care professionals and the public.* Cassels discusses recent Canadian focus group data that reveal a lack of knowledge about nutrigenomics among the general public and health care practitioners. These focus group participants strongly preferred government to provide information to the public about nutrigenomics and to ensure oversight of professionals and companies involved in service delivery. Saukko's chapter, *Pitching products, pitching ethics: selling nutrigenetic tests as lifestyle or medicine,* discusses online marketing and how companies characterize tests as either oriented toward medicine or lifestyle. She also comments on how regulation influences companies' representations about their services and products. Caulfield, Shelley, Bubela and Minaker's chapter, *Framing nutrigenomics for individual and public health: public representations of an emerging field,* examines various claims about public health benefits of nutrigenomics and cautions that barriers in cost and availability of healthy foods may undermine achievement of these benefits.

Nutrigenomics companies center their product and service offerings on personalized testing and dietary advice. This focus has raised questions about the accessibility of nutrigenomics and potential for nutrigenomics to be offered to the public. In *The personal and the public in nutrigenomics,* Castle explores the issue of how nutrigenomics has followed a path toward personalization and investigates whether public health applications, not just testing services focused on individuals, of nutrigenomic science are possible. Korthals takes this issue even further, in his reflection on *Food styles and the future of nutrigenomics.* He asks whether nutrigenomics must be configured as a science in pursuit of health through diet or whether there are other options for nutrigenomics, such as derivation of greater sensory pleasure from eating foods for which there is a genomic basis for taste preference.

The chapters in this book follow nutrigenomics from the laboratory bench, to the market, regulatory domains, individuals and health care professionals, the Internet and other media, to the public and political realm. The growth in nutrigenomics science and its applications in the marketplace and health systems raise a variety of ethical, legal and social issues that have been identified in existing literature, but not necessarily analyzed with a view to proposing strategies for addressing the challenges. This book attempts to move the discussion in this direction by developing a set of constructive responses to major issues the authors identify in science, ethics, law, regulation and communication. Authors were asked to describe the three most important issues raised in their chapter and to propose how the issues could be addressed through research, business activity, regulation, communication, policy setting or political action. The purpose of the *Epilogue* is to crystallize the content of the book in a set of recommendations about the most important issues in the current cycle of nutrigenomics to provide solutions to current problems and to prepare for subsequent developments in the field.

We conclude this introduction with a few words about the front and back cover photographs, which were taken, respectively, by Tamara Rodela and David Castle. The front photograph was taken in an urban fruit and vegetable market in Québec City. At the time of writing, Québec City just celebrated the 400th anniversary of its founding. In the 1600s, settled European life expectancy was just under 50 years – up an entire decade from 1500 due to improved nutrition and sanitation. When Samuel de Champlain arrived on July 3, 1608 to settle the abandoned fort built by Jacques Cartier in 1535, there was no welcome basket waiting for him with the fruits and vegetables pictured on the front cover. In fact, only eight of the 28 of his ship's crew survived the first winter's harsh conditions. Now a UNESCO World Heritage site, the walled city and buildings of the old port of Québec City may be beautifully preserved for contemporary residents and visitors to enjoy, but fortunately massive improvements have been made to de Champlain's nutritional environment.

The back photograph was taken at a seasonal farmer's market on Mayne Island, a rural island of 900 residents located in Canada's Pacific Northwest an hour away by ferry from Vancouver. Productive fruit orchards and vegetable greenhouses thrived on the island from the late 1800s to the mid-1900s. Today, small-scale farmers work hard to keep local agriculture alive on Mayne and neighboring islands, but they have a dedicated following among market-goers who are committed to the nutritional value and health benefits of local, organically produced food. The hand-lettered sign in the photograph appeared in the center of the market, ringed by produce stalls, locally prepared foods and tables offering arts and crafts. Island residents have access to a doctor, but only a few days out of every week and rely on a local courier service to take their blood samples to the city for medical tests and return with any required prescription medication.

An urban dweller in contemporary Québec City, steeped in all the advances of modern life, seems a likely candidate to be interested in nutrigenomics. Yet there is ample evidence that people in small communities such as Mayne Island are similarly concerned with the connections between food and health. As a growing area of science and technology, nutrigenomics has relevance for anyone who is interested in relationships between the food we eat, our genetics and our health.

CHAPTER

1

Gene–Environment Interactions: Where are we and where should we be Going?

Jose M. Ordovas and E. Shyong Tai

OUTLINE

Nutrition and Genomics
ISBN: 978-0-12-374125-7

SUMMARY

This chapter examines the reasons for studying gene–environment interactions and evaluates recent reports of interactions between genes and environmental modulators in relation to cardiovascular disease and its common risk factors. Recent studies focusing on smoking, alcohol, physical activity and coffee are all observational and include relatively large sample sizes. They tend to examine a single gene and fail to address interactions with other genes as well as other correlated environmental factors. Studies examining gene–diet interactions include both observational and interventional designs, and are of smaller scale, especially those including dietary interventions. Among the reported gene–diet interactions, it is important to highlight the strengthening evidence for APOA5 as a major gene involved in triglyceride metabolism and modulated by dietary factors and the identification of APOA2 as a modulator of food intake and obesity risk. This chapter concludes that, overall, the study of gene–environment interactions is an active and much needed area of research. While technical barriers in genetic studies are being quickly overcome, the inclusion of comprehensive and reliable environmental information represents a significant shortcoming to genetic studies. Progress depends on larger study populations being included and also more comprehensive, standardized and precise approaches to capture environmental information.

INTRODUCTION

After more than two decades of great expectations but few deliverables, the field of genetics related to common and complex disorders has made remarkable progress towards the identification of novel loci and genetic variants associated with these diseases. This has been possible thanks to the combination of more robust experimental approaches, including large population studies, with the availability of high density genotyping (>1 million single nucleotide polymorphisms (SNPs)). At the time of writing, there have been both consolidation of some of the traditional candidate genes, especially in the area of lipid metabolism, and, more exciting, identification of new lipid-related loci. These findings will provide us with a more complete understanding of the metabolic landscape and new insights into the pathogenesis of disease (Smith, 2007; Zeggini and McCarthy, 2007). The search is far from over, however, and for the newly discovered and for well-known candidate genes, current knowledge needs to be augmented by using deep re-sequencing and phenotyping of individuals carrying functional variants at these loci to understand the pathways and metabolic fluxes affected by these genetic variants and to provide greater insight into the physiologic basis of disease (Tracy, 2008). Based on current knowledge, however, many of the observed gene effects will not be insulated from environmental modulation. Therefore, there is a compelling need to further the initial association studies with well-designed investigation of gene–environment interactions.

GENES OR ENVIRONMENT? LIMITATIONS OF THE TRADITIONAL APPROACH TO DISEASE RISK ASSESSMENT

From an epidemiologic perspective, studies of genetic and environmental factors will continue to underestimate the population-attributable risk associated with either genetic or environmental factors. In fact, consideration of the joint effects of genetic and environmental factors strengthens their associations with disease, permitting the identification of risk factors that have small marginal effects. Even the best well established genetic markers for common traits show inter-population differences. For example, the recently identified fat mass and obesity-associated (FTO) gene has been heralded as the most solid locus for obesity risk. Yet there are conflicting reports of lack of association in African Americans or Han Chinese (Li et al., 2008). It remains unclear whether these are due to genetic differences between the different populations, or whether a gene–environment interaction may be masking the effect in these other ethnic groups. Findings from a Danish population, for example, indicate that physical activity may attenuate the effects of the FTO genetic variants in support of gene–environment interactions (Andreasen et al., 2008).

Reliably capturing environmental measures and the complexity of the dietary 'environment' present major difficulties in studying gene–environment interactions. Methods to capture dietary information from observational studies have been criticized for years for a lack of accuracy, precision and objectivity. Moreover, foods are very complicated mixtures and we may attribute an observed effect to a specific nutrient because we know about it but, in reality, it may be due to another component of the food that we may not know or pay attention to. For example, coffee consumption shows a well-replicated association with the risk of type 2 diabetes mellitus (van Dam and Hu, 2005; Pereira et al., 2006). But coffee is a complex mixture of compounds and it is common to equate coffee with caffeine. Still, the specific compound in coffee that produces this effect is unclear. In fact, caffeine may not be important after all, since the association is also seen with decaffeinated coffee (van Dam et al., 2006). Therefore, integration of genetic variability with gene expression studies will pinpoint pathways that the environmental exposures may act through and will provide some guidance as to the specific environmental components that warrant further investigation.

From a public health perspective, it has been suggested that the identification of genetic variants that encode susceptibility to disease could be used in risk algorithms to identify individuals at high risk of disease. The identification of gene–environment interactions may suggest specific interventions that may attenuate the risk in these individuals. Taken to an extreme, it has been suggested that these studies could lead to individualized health plans based on a person's genetic make-up (Subbiah, 2007). In this chapter, we summarize current knowledge related to gene–environment interactions as they pertain to cardiovascular disease risk factors, particularly those related to metabolic phenotypes. In addition, we raise some of the issues that need to be addressed to advance this field of research and to develop applications for the public.

CARDIOVASCULAR RISK FACTORS

The scientific literature is heavily populated with papers reporting cardiovascular risk factors. The latest catalog published over one decade ago listed 177 of them with many of them falling in the categories of 'Nutrition-Related' and 'Environmental' (Omura et al., 1996). Many of the reported risk factors are rather questionable, however. Updating this catalog today would probably add a few hundred more to the list. Besides diet, the best characterized environmental risk factors include smoking, inadequate physical activity, alcohol and, more recently, coffee drinking (which, given its popularity and widespread use, has received increased attention as a modifier of cardiovascular disease risk) (Campos and Baylin, 2007). Those common and well established behavioral risk factors are the focus of this section.

Gene–Smoking Interactions

The study of interactions between genetic factors and smoking has been an active area of research primarily in the fields of cancer and neurodegenerative diseases (Wang and Wang, 2005; Elbaz et al., 2007), but it also caught the early attention of cardiovascular researchers (Kondo et al., 1989) and a substantial body of evidence has accumulated during the last two decades, as summarized by recent reviews (Talmud, 2007). Considering what we know about the risk associated with tobacco smoking, it is obvious that all the reports are supported by observational data and there are no randomized intervention studies. Of the behavioral factors contemplated in this section, tobacco smoking may be the most reliable in terms of validity of reporting. Moreover, it is a variable most epidemiological studies collect. Therefore, studies reporting gene–tobacco smoking interactions tend to be large and probably have adequate statistical power to examine single gene–single factor interactions. This is still far from ideal as no single study can yet fully address the complex interactions involving multiple genes and environmental factors. Genes currently under examination span a variety of metabolic pathways, including the obvious ones involved in drug metabolism and detoxification, those involved in lipid traits known to be modified by smoking (Gambier et al., 2006; Goldenberg et al., 2007; Cornelis et al., 2007b; Manfredi et al., 2007) as well as others involved in a variety of more remotely connected metabolic functions (Lee et al., 2006; Saijo et al., 2007; Jang et al., 2007; Stephens et al., 2008). In addition, the arrival of genome-wide association studies has allowed the identification of chromosomal regions of interest (North et al., 2007) for which no candidate genes have yet been identified. The increased size of recent studies and the creation of large consortia are allowing the examination of disease events and non-invasive biomarkers of disease (Lee et al., 2006; Goldenberg et al., 2007; Cornelis et al., 2007b; Manfredi et al., 2007; North et al., 2007; Stephens et al., 2008) (Table 1.1). As expected from the publication bias, most published reports identify significant gene–smoking interactions and most of the studies, but not all of them (Goldenberg et al., 2007; Cornelis et al., 2007b; Saijo et al., 2007), conclude that carriers of the minor alleles were more susceptible to the deleterious effects of tobacco smoking. While uncovering gene–smoking interactions may reveal interesting physiological mechanisms, the practical utility of this knowledge is limited as the relevant public health advice is that every smoker should quit, regardless of genotype.

Gene–Alcohol Interactions

As with gene–smoking interactions, lipid researchers have long been interested in interactions between genetic factors and alcohol consumption (Hannuksela et al., 1994). Like smoking, recent reports describing gene–alcohol interactions are based on observational studies with information about alcohol drinking derived from medical records or questionnaires involving thousands of subjects (Beulens et al., 2007; Jensen et al., 2008; Volcik et al., 2007; Tsujita et al., 2007) (Table 1.2). The genes selected include traditional lipid candidate genes (Tsujita et al., 2007; Jensen et al., 2008; Volcik et al., 2007) as well as those involved in alcohol metabolism (Beulens et al., 2007). For the most part, the modulating effect appears to be driven by HDL-C concentrations as reported in earlier work (Hannuksela et al., 1994). Moderate alcohol consumption is positively associated with health, but the potential benefits may not be universal. Knowledge of these studies could help identify subjects who benefit from moderate drinking, those who have no effect and those who experience harm.

TABLE 1.1 Gene–smoking interactions

Gene	SNP	Population	Trait	Outcome	Reference
GSTM1, GSTT1 and CYP1A1	CYP1A1 (MspI); GSTM1(null) and GSTT1(null)	222 consecutive smoker patients	CAD	Smokers carrying GST deleted genotypes have an increased susceptibility to smoking related CAD	Manfredi et al., 2007
IL6	-634C/G	347 men	CRP	The impact of the -634G allele on CRP elevation is greater in non-smokers than in current smokers.	Saijo et al., 2007
LTA	252A > G	480 men	Lipids, glucose, TNF-alpha, IL-6, CRP, adiponectin and 8-epi PGF2alpha	The LTA 252A > G polymorphism may modulate the inflammatory effects and oxidative stress of smoking. The detrimental effect of smoking is most clearly seen in men with G/G genotype	Jang et al., 2007
Chromosomes 4, 6, 11 and 13	D4S2297, D6S1056, D11S199, D13S892	2128 men and women	CCP	Additive and QTL-specific genotype-by-smoking interaction was detected on chromosomes 4, 6, 11 and 13	North et al., 2007
CETP	TaqIB	199 current smokers; 345 past smokers; 270 never smokers	First MI	Smoking was a risk factor for B1B1 and B1B2 subjects, but not so for the B2B2 carriers	Goldenberg et al., 2007
GSTM1, GSTT1 and GSTP1	del SNPs in GSTM1 and GSTT1 and Ile105Val in GSTP1	2042 first acute non-fatal MI cases and 2042 controls	MI	Consumption of cruciferous vegetables was associated with a lower risk of MI among those with a functional GSTT1*1 allele	Cornelis et al., 2007b

UCP2	-866G > A	453 diabetic men	CAD	This study demonstrates an interaction between the UCP2 -866G > A variant and smoking to increase oxidative stress *in vivo*	Stephens et al., 2008
CYP1A1 and AHR	CYP1A1 T3801C and AHR G1661A	302 men and 311 women	Blood pressure	Interaction between the CYP1A1 T3801C and AHR G1661A SNPs, smoking and blood pressure level	Gambier et al., 2006
NOS3	T-786C and E298D	1085 CHD and 300 ischemic stroke cases, and 1065 controls	CVD	Interaction between the E298D and T-786C polymorphisms in NOS3, cigarette smoking, and risk of CHD and ischemic stroke events	Lee et al., 2006

TABLE 1.2 Gene– (alcohol/physical activity/coffee) interactions

Gene	SNP	Population	Trait	Factor	Outcome	Reference
ADH1C	ADH1C*1/*2	640 diabetic and 1000 control women; 383 diabetic and 382 control men	T2DM	Alcohol	The ADH1C*2 allele, related to a slower oxidation rate, attenuates the lower diabetes risk among drinkers	Beulens et al., 2007
CETP	TaqIB	659 men and 1070 women	HDLC	Alcohol	The association between the TaqIB SNP and HDL-C levels was more evident in drinkers than in non-drinkers	Hannuksela et al., 1994
CETP	TaqIB	1504 men and women CAD cases and controls	HDLC and CHD	Alcohol	The association of alcohol with HDL-C levels was modified by the TaqIB genotype and there was a suggestion of interaction on CHD risk	Jensen et al., 2008
CETP, LPL, HL, PON1	CETP TaqIB; LPL Ser447X; LIPC -514C/T; PON1Q192R	8772 men and women	CHD	Alcohol	Interactions between alcohol and candidate genes for HDL-C metabolism influence the risk of incident CHD in black men	Volcik et al., 2007
ABCC8 and SLC2A2	ABCC8 (rs3758947) and SLC2A2 (rs5393, rs5394, rs5404)	479 overweight subjects with IGT	T2DM	Physical activity	Moderate-to-vigorous PA may modify the risk of developing T2D associated with genes regulating insulin secretion (SLC2A2, ABCC8) in persons with IGT	Kilpelainen et al., 2007
EDN1	rs2070699, rs5369, rs5370, rs4714383, and rs9296345	1956 men and women (cases and controls)	Blood pressure	Physical activity	The expression of the genotype effect is modulated by physical activity or cardiorespiratory fitness level	Rankinen et al., 2007
GNB3	825C > T	14716 mean and women	Obesity and hypertension	Physical activity	Potential interaction between the GNB3 gene, physical activity and obesity status which may influence hypertension prevalence in African Americans	Grove et al., 2007

PPARG	Pro12Ala	1481 men and women	T2DM	Physical activity	There may be a gene–environment interaction between the Pro12 allele of the PPAR-gamma gene and physical activity that results in increased risk of T2DM in Non-Hispanic Whites	Nelson et al., 2007a
FTO	rs9939609	17508 men and women	BMI and T2DM	Physical activity	The FTO gene was associated with T2DM when not adjusted for BMI. Moreover, physical activity accentuates the effect of FTO on body fat accumulation	Andreasen et al., 2008
ADORA2A and CYP1A2	ADORA2A (1083C/T) and CYP1A2 (-163A/C)	2735 men and women, CAD cases and controls	Caffeine consumption	Coffee	Probability of having the ADORA2A 1083TT genotype decreases as habitual caffeine consumption increases. No association with the CYP1A2 gene	Cornelis et al., 2007a

Gene–Physical Activity Interactions

Physical inactivity increases the relative risk of most common diseases plaguing industrialized societies including coronary artery disease, stroke, hypertension, osteoporosis and cancer (Booth and Lees, 2007). For thousands of years, humans had to be active to survive and metabolic pathways selected during the evolution of the human genome are inevitably linked to physical activity. Sedentary modern lifestyles may prompt a maladaptive response to our ancestral genomic background that leads to metabolic dysfunction and many chronic diseases. The topic of gene–physical activity interactions is receiving increasing attention (Andreasen et al., 2008; Grove et al., 2007; Kilpelainen et al., 2007; Rankinen et al., 2007; Nelson et al., 2007a) (see Table 1.2). Similar to smoking and alcohol, all recent reports are based on relatively large observational studies. The selection of genes studied appears to be quite heterodox, with no clear unifying theme and also with a variety of main outcomes including diabetes (Kilpelainen et al., 2007; Nelson et al., 2007a), blood pressure (Rankinen et al., 2007; Grove et al., 2007) and obesity (Grove et al., 2007). The downside of the current research is that some of the studies were not initially intended to focus on physical activity and the quality of the information may not meet the standards of precision and accuracy needed to perform reliable interaction analyses.

Gene–Diet Interactions

Most of the emphasis on gene–environment interactions continues to be placed on gene–diet interactions. Unlike studies focusing on tobacco smoking, alcohol and physical activity, studies of gene–diet interactions include both observational and intervention studies. The major concern with extracting information about dietary intake from observational studies continues to be the use of dietary intake assessments that do not accurately reflect true intake of the individual. Improvements in data collection are needed to obtain better and more objective measures of nutrient intakes from large observational studies. Conversely, the major concern about intervention studies has been, and continues to be, the very small number of subjects involved in each study (Ordovas and Corella, 2004; Corella and Ordovas, 2005). For the purpose of this chapter, we have grouped studies according to their experimental design and include postprandial studies that were not accompanied by dietary changes within observational studies. Previous reports have extensively reviewed this topic (Ordovas and Corella, 2004; Corella and Ordovas, 2005) and in this work we will focus only on recent publications.

Observational studies reporting gene–diet interactions in the fasting or postprandial states overwhelmingly focused on the traditional lipid candidate genes (Santos et al., 2006; Corella et al., 2007b; Robitaille et al., 2007b; Li et al., 2007; Morcillo et al., 2007; Shen et al., 2007; Sofi et al., 2007; Scacchi et al., 2007) (Table 1.3). The average number of subjects included in observational gene–diet interaction studies is much lower than those reported for interactions with alcohol, smoking and physical activity and these numbers have not changed much in recent years. The main outcomes examined in gene–diet interactions have included primarily plasma lipids, but also body mass index (BMI), inflammatory markers and other measures of the metabolic syndrome. Moreover, each report is limited to one single locus and, even those that examine multiple loci (Santos et al., 2006) do not attempt to examine more complex but more realistic situations involving gene–gene–nutrient interactions.

It is important to underscore the consolidation of certain candidate genes on their role in lipid metabolism and modulation by dietary factors. This is the case of the APOA5 gene (Corella et al., 2007b; Hubacek et al., 2007; Moreno-Luna et al., 2007), as well as new findings involving the long-known but cryptic APOA2, suggesting potential roles of this apolipoprotein on dietary intake, body mass index and postprandial lipemia (Corella et al., 2007a; Delgado-Lista et al., 2007).

Although most of the gene–diet interactions have focused on dietary fats, other habits such as coffee drinking have been studied in a recent case-control study (Cornelis et al., 2007a) that examined the interaction between the adenosine A2A receptor (ADORA2A) and the CYP1A2 genes and caffeine intake (see Table 1.2) and reported a modulation of caffeine intake due to genetic variability at the ADORA2A gene. Several other phenotypes have been investigated in relation to gene–diet interactions beyond dietary fats. One of the most solidly established is the 5′,10′-methylenetetrahydrofolate reductase (MTHFR) gene which has been comprehensively reviewed (Cummings and Kavlock, 2004; Friso and Choi, 2005). Another less explored example is the one provided by investigators in the UK and New Zealand who tested whether children's intellectual development was influenced by both genetics and early nutrition – i.e. breastfeeding versus formula feeding (Caspi et al., 2007). These researchers showed that the association between breastfeeding and IQ is moderated by a genetic variant in FADS2, a gene involved in the genetic control of fatty acid pathways. Their finding shows that environmental exposures can be used to uncover novel candidate genes in complex phenotypes. It also shows that genes interact with the early environment to shape the complex phenotypes such as IQ.

Findings from observational studies are informative and are of great potential value, but their clinical validity must be confirmed by intervention studies. Ideally, studies designed to test gene–diet interactions should involve prior selection of participants based on the genotype of interest. Most of the currently reported intervention studies are still

TABLE 1.3 Gene–diet interactions (observational studies)

Gene	SNP	Population	Main trait	Outcome	Reference
SLC6A14, CART, GHSR, GAD2, GHRL, MKKS, LEPROTL1, PCSK1, UCP2, UCP3, FOXC2, PPARGC1A, PPARG2, PPARG3, SREBF1, WAC, HSD11B1, LIPC, IGF2, KCNJ11, ENPP1, ADIPOQ, CD36, IL6, TNFA, SERPINE1	43 SNPs	549 obese women	Obesity	The most remarkable interaction found in this study refers to the combination of the hepatic lipase gene polymorphism -514 C > T and fiber intake	Santos et al., 2006
APOA1/C3/A4/A5	APOA1 (G-75 > A and C83 > T), APOC3 (C-482 > T and C3238 > G), APOA4 (Thr347 > Ser and Gln360His) and APOA5 (T-1131 > C, Ser19 > Trp and Val153 > Met)	133 men	Lipids	APOA4 and APOA5 variants may play an important role in the individual sensitivity of lipid parameters to dietary composition in men	Hubacek et al., 2007
APOA5	APOA5-1131T > C and 56C > G (S19W)	1073 men and 1207 women	BMI	The APOA5-1131T > C modulates the effect of fat intake on BMI and obesity risk in both men and women	Corella et al., 2007b
CPT1	CPT1B c.282-18C > T and p.E531K variants	252 men and 99 women	Obesity	The findings suggest that indices of obesity might be modulated by an interaction between CPT1 variants and fat intake	Robitaille et al., 2007b

CETP	TaqIB	780 diabetic men	HDLC	These data suggested an interaction between the CETP TaqIB polymorphism and intake of dietary fat on plasma HDL concentration	Li et al., 2007
FABP2	Ala54Thr	1226 men and women	Insulin resistance	An interaction existed between the Ala54Thr polymorphism of FABP2 and the intake of dietary fats in a population with a high intake of monounsaturated fatty acids	Morcillo et al., 2007
IL1B	-1473G > C, -511G > A, -31T > C, 3966C > T, 6054G > A	540 men and 580 women	Inflammation and the MetS	These results suggested that IL1beta genetic variants were associated with measures of chronic inflammation and the MetS risk and that genetic influences were more evident among subjects with low (n-3) PUFA intake	Shen et al., 2007
LPA	LPA 93C > T and LPA 121G > A	260 men and 387 women	Lp(a)	This study reported a significant interaction of daily fish intake and LPA 93C > T polymorphism in decreasing Lp(a) concentrations	Sofi et al., 2007
PPARG	Pro12Ala	1394 men and women	T2DM	This analysis suggests that the protective effect of the Ala allele against T2DM is present only in populations with a diet rich in lipids	Scacchi et al., 2007

TABLE 1.3 (*Continued*)

Gene	SNP	Population	Main trait	Outcome	Reference
APOA2	m265T > C	514 men and 564 women	BMI	The -265T > C polymorphism is consistently associated with food consumption and obesity, suggesting a new role for APOA2 in regulating dietary intake	Corella et al., 2007a
APOA2	m265T > C	88 E3/E3 men	PPL	This work shows that carriers of the minor C allele for APOA2 -265T/C (CC/TC) have a lower postprandial response compared with TT homozygotes	Delgado-Lista et al., 2007
APOA5	APOA5-1131T > C and 56C > G (S19W)	88 E3/E3 men	PPL	The minor 56G and -1131C alleles were associated with a higher postprandial response	Moreno-Luna et al., 2007
PPARA	Leu162Val and 140+5435T > C	59 men	Fasting and PPL	These data suggest that PPARalpha variants may modulate CVD risk by influencing both fasting and postprandial lipid concentrations	Tanaka et al., 2007b
APOB	m516C/T	47 E3/E3 subjects	PPL	The presence of the C/T genotype is associated with a higher PPL response	Perez-Martinez et al., 2007b
SCARB1	c1119C > T	59 men	PPL	The c.1119C > T polymorphism is associated with a lower postprandial TG response in the smaller, partially catabolized lipoprotein fraction	Tanaka et al., 2007a

using retrospective and opportunistic analyses of subjects participating in dietary intervention studies designed for non-genetic purposes (Table 1.4) (Moreno-Luna et al., 2007; Perez-Martinez et al., 2007a, 2007b, 2007c; Fernandez de la Puebla et al., 2007; Tanaka et al., 2007a, 2007b; Delgado-Lista et al., 2007; Santosa et al., 2007; Hubacek et al., 2007; Nelson et al., 2007b; Corella et al., 2007a; Lai et al., 2007; Erkkila et al., 2007; Weiss et al., 2007; Helwig et al., 2007; Robitaille et al., 2007a, 2007b; Nieminen et al., 2007). As expected from their cost and complexity, the number of participants in these studies is very small and subject to errors and spurious associations and interactions. Similar to observational studies, most of the interventional reports focused on well-known candidate genes and none of the novel loci identified from genome-wide association studies has yet made their way to the literature. Although it is anticipated that reports about new genes will trickle into the literature in the near future.

STATISTICAL METHODS FOR GENE–ENVIRONMENT INTERACTIONS

In addition to issues related with experimental designs, the statistical approach to the analyses of gene–environment interaction has emerged as another significant downstream problem with these studies (Boks et al., 2007; Chanda et al., 2007; Zhang et al., 2008; Bureau et al., 2008). It is well known that large sample sizes are essential for the consolidation of associations of new loci with common disorders and traits. Sample size becomes even more critical when it comes to analyzing gene–environment interactions. Unfortunately, several features related to the study of environmental exposures further complicate the analyses and interpretation of interactions.

The current limitations no longer come from the now very precise process of ascertaining genotype. This means that measurement error does not play a large role in biasing the estimates of risk associated with a genetic variant. In fact, most measurement error in a genome-wide association study comes from the fact that not every variant is studied but a set of genetic variants is utilized as proxies for functional variants. In this regard, the technological advances that allow the inclusion of greater numbers of variants on a single array, as well as the careful selection of variants, means that reasonable power is available for the detection of most common variants across the genome (Teo, 2008). In contrast, the assessment of environmental exposures through diet is much less precise. Genetic analysis is also remarkably stable across genotyping platforms with 95% or greater concordance across most platforms (Servin and Stephens, 2007; Marchini et al., 2007). This makes meta-analyses (whether *post-hoc* or *a priori*) relatively

TABLE 1.4 Gene–diet interactions (intervention studies)

Gene	SNP	Population	Trait	Factor	Outcome	Reference
APOB	APOB -516C/T	30 men and 29 women	Insulin sensitivity	Dietary fat	Male carriers of the -516T allele, C/T, have a significant increase in insulin resistance after consumption of all diets, but the difference is more exaggerated after the SFA diet compared with the MUFA- and CHO-rich diets	Perez-Martinez et al., 2007c
APOB	APOB -516C/T	70 men and 27 women	Lipids	Dietary fat	Data suggest that the APOB -516C/T SNP has no effect on the lipid profile after changes in dietary fat in a healthy population	Perez-Martinez et al., 2007a
F7	R353Q and 5′F7	30 men and 29 women	FVII Ag levels	Dietary fat	Data show that carriers of the RR and/or A1A1 genotype present higher FVII Ag levels after the consumption of a SAT diet compared with the MEDIT and CHO rich diets	Fernandez de la Puebla et al., 2007
ABCG5 and ABCG8	Q604E and C54Y	35 hypercholes-terolemic women	Weight loss	Hypocaloric diet and physical activity	SNPs in ABCG5/G8 were found to be associated with the response of cholesterol metabolism to weight loss	Santosa et al., 2007
ADIPOQ	276 and 45	57 men and women	Adiponectin	Dietary ALA	ALA supplementation for 8 weeks may lower adiponectin levels among healthy individuals independent of SNPs in the adiponectin gene	Nelson et al., 2007b
APOA5	APOA5-1131T > C and 56C > G (S19W)	791 men and women	PPL	Fat load and fenofibrate	APOA5 56G carriers benefited more from the fenofibrate treatment than non-carriers in lowering plasma TG and increasing HDL-C levels	Lai et al., 2007

APOE	APOE2	34 men and women	Lipids	Dietary sucrose	Moderate increase in sucrose intake does not affect fasting or postprandial serum lipid responses in healthy subjects with or without the E2 allele	Erkkila et al., 2007
FABP2	Ala54Thr	122 men and women	Glucose tolerance	Dietary fat	Following a low-fat diet, FABP2 Thr54 carriers have lower glucose tolerance and lower insulin action than do Ala54-homozygotes. This may be due to higher lipid oxidation rates	Weiss et al., 2007
FABP2	Ala54Thr and Promoter SNP	700 men	TG and HOMA	Diet (post-prandial load)	Higher postprandial TG levels and a decrease in insulin sensitivity in T54T subjects were expressed only in the presence of the BB genotype at the promoter SNP. Similar results were obtained after oral glucose tolerance test. This is probably due to a higher responsiveness to PPARG/RXR of the FABP2 promoter B vs promoter A	Helwig et al., 2007
LXRA	c.-115G > A, c.-840C > A and c.-1830T > C	35 men	Lipids	Diet (cholesterol intake)	These results suggest that cholesterol intake interacts with LXRA variants to modulate the plasma lipid profile.	Robitaille et al., 2007a
SCAP and APOE	SCAP (Ile796Val) and apo E	78 women	Weight loss	Very low energy diet	Neither the SCAP nor the APOE SNPs were associated with weight loss. However, the APOE genotype seems to be one of the modifying factors for cholesterol concentrations during very low-energy diet	Nieminen et al., 2007

straightforward. Quite the opposite, there are significant differences in methodologies for the assessment of environmental exposures, particularly for more complex exposures such as dietary intake.

In addition, environmental exposures often show significant colinearity. For example, individuals who eat a healthy diet may also participate in other health promoting activities. Yet, studies that focus on nutritional aspects have limited information, if any, about other variables such as physical activity and vice versa. For the most part, this is a consequence of sample size limitations. Failure to take these colinearities into account could result in severe confounding and misinterpretation. This is less of a problem for genetic associations studies in the era of the genome-wide association study since most of the colinearity in the human genome (represented by the patterns of linkage disequilibrium within the human genome) can be accounted for.

It is therefore crucial to develop tools for the assessment of environmental exposures that provide more precise estimates of exposure and that can be readily standardized across studies. The development of a validation for tools to assist in dietary recall (Bingham, 2002; Williamson et al., 2003; Baylin and Campos, 2006) and the use of biomarkers to assess nutrient intake represent important steps towards achieving this goal. In addition, consensus is needed about what environmental factors should be measured. The added complexity of unifying the information about environment has made much more difficult the adoption of similar approaches as those currently used for the study of associations (*a priori* meta-analysis). Studies continue to be dramatically underpowered, specifically those related to gene–diet interactions where the capture of the information is more complex than for other traits such as smoking.

PERSONALIZED NUTRITION AND THE CONSUMER

The pursuit of research on gene–environment interactions has the primary goal of gaining a better understanding of the impact of nature and nurture on an individual's metabolism. The ultimate goal is to translate this knowledge into practical applications that guide the individual in optimizing health. Some consumers are interested in health benefits of personalizing nutrition and other interventions and this has resulted in some genetic testing products being offered to provide 'personal dietary advice' for consumers. Two recent publications (Kornman et al., 2007; Arkadianos et al., 2007) provide some support for benefits of individualized genetic advice. Yet, the studies are very limited in size and tests offered have to be further validated and considerably improved to have a significant impact on individual and population health.

SUMMARY

The number of publications addressing gene–environment interactions reflects a robust interest towards gaining much needed knowledge in the area of metabolic traits and cardiovascular diseases. The approaches in use have not kept up with the real needs of the field, however. Most studies, and especially those focusing on gene–diet interactions, are carried out with less than optimal sample sizes. Further studies on gene–environment interactions should follow recent steps of simple genotype–phenotype association studies and carry out *a priori* meta-analyses. Standardization of the 'environment' is, however, much more complex than the standardization of clinical or biochemical traits and more emphasis should be placed on new and standardized methods to capture environmental information. In the meantime, new statistical methods are needed to deal with the current constraints. Despite the current uncertainties and limitations, the concept of gene–environment interactions modulating common disease risk factors is well founded and should provide future scientific knowledge to address major health problems using molecular and more individually targeted approaches to disease prevention and therapy.

ACKNOWLEDGMENT

The authors acknowledge support from the National Institutes of Health, National Institute on Aging, Grant number 5P01AG023394-02 and NIH/NHLBI grant numbers HL54776 and U01HL072524-04 and NIH/NIDDK DK075030 and contracts 53-K06–5-10 and 58–1950-9–001 from the US Department of Agriculture Research Service.

References

Andreasen, C.H., Stender-Petersen, K.L., Mogensen, M.S., et al. (2008). Low physical activity accentuates the effect of the FTO rs9939609 polymorphism on body fat accumulation. *Diabetes* 57:95–101.

Arkadianos, I., Valdes, A.M., Marinos, E., Florou, A., Gill, R.D. and Grimaldi, K.A. (2007). Improved weight management using genetic information to personalize a calorie controlled diet. *Nutr J* 6:29.

Baylin, A. and Campos, H. (2006). The use of fatty acid biomarkers to reflect dietary intake. *Curr Opin Lipidol* 17:22–27.

Beulens, J.W., Rimm, E.B., Hendriks, H.F., et al. (2007). Alcohol consumption and type 2 diabetes: influence of genetic variation in alcohol dehydrogenase. *Diabetes* 56: 2388–94.

Bingham, S.A. (2002). Biomarkers in nutritional epidemiology. *Public Health Nutr* 5:821–27.

Boks, M.P., Schipper, M., Schubart, C.D., Sommer, I.E., Kahn, R.S. and Ophoff, R.A. (2007). Investigating gene environment interaction in complex diseases: increasing power by selective sampling for environmental exposure. *Int J Epidemiol* 36:1363–9.

Booth, F.W. and Lees, S.J. (2007). Fundamental questions about genes, inactivity, and chronic diseases. *Physiol Genomics* 28:146–57.

Bureau, A., Diallo, M.S., Ordovas, J.M. and Cupples, L.A. (2008). Estimating interaction between genetic and environmental risk factors: efficiency of sampling designs within a cohort. *Epidemiology* 19:83–93.

Campos, H. and Baylin, A. (2007). Coffee consumption and risk of type 2 diabetes and heart disease. *Nutr Rev* 65:173–79.

Caspi, A., Williams, B., Kim-Cohen, J., et al. (2007). Moderation of breastfeeding effects on the IQ by genetic variation in fatty acid metabolism. *Proc Natl Acad Sci USA* 104: 18860–5.

Chanda, P., Zhang, A., Brazeau, D., et al. (2007). Information-theoretic metrics for visualizing gene–environment interactions. *Am J Hum Genet* 81:939–63.

Corella, D. and Ordovas, J.M. (2005). Single nucleotide polymorphisms that influence lipid metabolism: interaction with dietary factors. *Annu Rev Nutr* 25:341–90.

Corella, D., Arnett, D.K., Tsai, M.Y., et al. (2007a). The -256T>C polymorphism in the apolipoprotein A-II gene promoter is associated with body mass index and food intake in the genetics of lipid lowering drugs and diet network study. *Clin Chem* 53: 1144–52.

Corella, D., Lai, C.Q., Demissie, S., et al. (2007b). APOA5 gene variation modulates the effects of dietary fat intake on body mass index and obesity risk in the Framingham Heart Study. *J Mol Med* 85:119–28.

Cornelis, M.C., El-Sohemy, A. and Campos, H. (2007a). Genetic polymorphism of the adenosine A2A receptor is associated with habitual caffeine consumption. *Am J Clin Nutr* 86:240–44.

Cornelis, M.C., El-Sohemy, A. and Campos, H. (2007b). GSTT1 genotype modifies the association between cruciferous vegetable intake and the risk of myocardial infarction. *Am J Clin Nutr* 86:752–58.

Cummings, A.M. and Kavlock, R.J. (2004). Gene–environment interactions: a review of effects on reproduction and development. *Crit Rev Toxicol* 34:461–85.

Delgado-Lista, J., Perez-Jimenez, F., Tanaka, T., et al. (2007). An apolipoprotein A-II polymorphism (-265T/C, rs5082) regulates postprandial response to a saturated fat overload in healthy men. *J Nutr* 137:2024–8.

Elbaz, A., Dufouil, C. and Alperovitch, A. (2007). Interaction between genes and environment in neurodegenerative diseases. *C R Biol* 330:318–28.

Erkkila, A.T., Schwab, U.S., Agren, J.J., Hallikainen, M., Gylling, H. and Uusitupa, M.I. (2007). Moderate increase in dietary sucrose does not influence fasting or postprandial serum lipids regardless of the presence of apolipoprotein E2 allele in healthy subjects. *Eur J Clin Nutr* 61:1094–101.

Fernandez de la Puebla, R.A., Perez-Martinez, P., Carmona, J., et al. (2007). Factor VII polymorphisms influence the plasma response to diets with different fat content, in a healthy Caucasian population. *Mol Nutr Food Res* 51:618–24.

Friso, S. and Choi, S.W. (2005). Gene-nutrient interactions in one-carbon metabolism. *Curr Drug Metab* 6:37–46.

Gambier, N., Marteau, J.B., Batt, A.M., et al. (2006). Interaction between CYP1A1 T3801C and AHR G1661A polymorphisms according to smoking status on blood pressure in the Stanislas cohort. *J Hypertens* 24:2199–205.

Goldenberg, I., Moss, A.J., Block, R., et al. (2007). Polymorphism in the cholesteryl ester transfer protein gene and the risk of early onset myocardial infarction among cigarette smokers. *Ann Noninvasive Electrocardiol* 12:364–74.

Grove, M.L., Morrison, A., Folsom, A.R., Boerwinkle, E., Hoelscher, D.M. and Bray, M.S. (2007). Gene–environment interaction and the GNB3 gene in the Atherosclerosis Risk in Communities study. *Int J Obes (Lond)* 31:919–26.

Hannuksela, M.L., Liinamaa, M.J., Kesaniemi, Y.A. and Savolainen, M.J. (1994). Relation of polymorphisms in the cholesteryl ester transfer protein gene to transfer protein activity and plasma lipoprotein levels in alcohol drinkers. *Atherosclerosis* 110:35–44.

Helwig, U., Rubin, D., Klapper, M., et al. (2007). The association of fatty acid-binding protein 2 A54T polymorphism with postprandial lipemia depends on promoter variability. *Metabolism* 56:723–31.

Hubacek, J.A., Bohuslavova, R., Skodova, Z., Pitha, J., Bobkova, D. and Poledne, R. (2007). Polymorphisms in the APOA1/C3/A4/A5 gene cluster and cholesterol responsiveness to dietary change. *Clin Chem Lab Med* 45:316–20.

Jang, Y., Koh, S.J., Kim, O.Y., et al. (2007). Effect of the 252A>G polymorphism of the lymphotoxin-alpha gene on inflammatory markers of response to cigarette smoking in Korean healthy men. *Clin Chim Acta* 377:221–27.

Jensen, M.K., Mukamal, K.J., Overvad, K. and Rimm, E.B. (2008). Alcohol consumption, TaqIB polymorphism of cholesteryl ester transfer protein, high-density lipoprotein cholesterol, and risk of coronary heart disease in men and women. *Eur Heart J* 29,1:104–12.

Kilpelainen, T.O., Lakka, T.A., Laaksonen, D.E., et al. (2007). Physical activity modifies the effect of SNPs in the SLC2A2 (GLUT2) and ABCC8 (SUR1) genes on the risk of developing type 2 diabetes. *Physiol Genomics* 31:264–72.

Kondo, I., Berg, K., Drayna, D. and Lawn, R. (1989). DNA polymorphism at the locus for human cholesteryl ester transfer protein (CETP) is associated with high density lipoprotein cholesterol and apolipoprotein levels. *Clin Genet* 35:49–56.

Kornman, K., Rogus, J., Roh-Schmidt, H., et al. (2007). Interleukin-1 genotype-selective inhibition of inflammatory mediators by a botanical, a nutrigenetics proof of concept. *Nutrition* 23:844–52.

Lai, C.Q., Arnett, D.K., Corella, D., et al. (2007). Fenofibrate effect on triglyceride and postprandial response of apolipoprotein A5 variants: the GOLDN study. *Arterioscler Thromb Vasc Biol* 27:1417–25.

Lee, C.R., North, K.E., Bray, M.S., et al. (2006). NOS3 polymorphisms, cigarette smoking, and cardiovascular disease risk: the Atherosclerosis Risk in Communities study. *Pharmacogenet Genomics* 16:891–99.

Li, H., Wu, Y., Loos, R.J., et al. (2008). Variants in the fat mass and obesity-associated (FTO) gene are not associated with obesity in a Chinese Han population. *Diabetes* 57: 264–68.

Li, T.Y., Zhang, C., Asselbergs, F.W., et al. (2007). Interaction between dietary fat intake and the cholesterol ester transfer protein TaqIB polymorphism in relation to HDL-cholesterol concentrations among US diabetic men. *Am J Clin Nutr* 86:1524–9.

Manfredi, S., Federici, C., Picano, E., Botto, N., Rizza, A. and Andreassi, M.G. (2007). GSTM1, GSTT1 and CYP1A1 detoxification gene polymorphisms and susceptibility to smoking-related coronary artery disease: a case-only study. *Mutat Res* 621:106–12.

Marchini, J., Howie, B., Myers, S., McVean, G. and Donnelly, P. (2007). A new multipoint method for genome-wide association studies by imputation of genotypes. *Nat Genet* 39:906–13.

Morcillo, S., Rojo-Martinez, G., Cardona, F., et al. (2007). Effect of the interaction between the fatty acid binding protein 2 gene Ala54Thr polymorphism and dietary fatty acids on peripheral insulin sensitivity: a cross-sectional study. *Am J Clin Nutr* 86:1232–7.

Moreno-Luna, R., Perez-Jimenez, F., Marin, C., et al. (2007). Two independent apolipoprotein A5 haplotypes modulate postprandial lipoprotein metabolism in a healthy Caucasian population. *J Clin Endocrinol Metab* 92:2280–5.

Nelson, T.L., Fingerlin, T.E., Moss, L.K., Barmada, M.M., Ferrell, R.E. and Norris, J.M. (2007a). Association of the peroxisome proliferator-activated receptor gamma gene with type 2 diabetes mellitus varies by physical activity among non-Hispanic whites from Colorado. *Metabolism* 56:388–93.

Nelson, T.L., Stevens, J.R. and Hickey, M.S. (2007b). Adiponectin levels are reduced, independent of polymorphisms in the adiponectin gene, after supplementation with alpha-linolenic acid among healthy adults. *Metabolism* 56:1209–15.

Nieminen, T., Matinheikki, J., Nenonen, A., et al. (2007). The relationship of sterol regulatory element-binding protein cleavage-activation protein and apolipoprotein E gene polymorphisms with metabolic changes during weight reduction. *Metabolism* 56: 876–80.

North, K.E., Carr, J.J., Borecki, I.B., et al. (2007). QTL-specific genotype-by-smoking interaction and burden of calcified coronary atherosclerosis: the NHLBI Family Heart Study. *Atherosclerosis* 193:11–19.

Omura, Y., Lee, A.Y., Beckman, S.L., Simon, R., et al. (1996). 177 cardiovascular risk factors, classified in 10 categories, to be considered in the prevention of cardiovascular diseases: an update of the original 1982 article containing 96 risk factors. *Acupunct Electrother Res* 21:21–76.

Ordovas, J.M. and Corella, D. (2004). Nutritional genomics. *Annu Rev Genomics Hum Genet* 5:71–118.

Pereira, M.A., Parker, E.D. and Folsom, A.R. (2006). Coffee consumption and risk of type 2 diabetes mellitus: an 11-year prospective study of 28 812 postmenopausal women. *Arch Intern Med* 166:1311–16.

Perez-Martinez, P., Perez-Jimenez, F., Ordovas, J.M., et al. (2007a). The APOB -516C/T polymorphism is associated with differences in insulin sensitivity in healthy males during the consumption of diets with different fat content. *Br J Nutr* 97:622–27.

Perez-Martinez, P., Perez-Jimenez, F., Ordovas, J.M., et al. (2007b). Postprandial lipemia is modified by the presence of the APOB-516C/T polymorphism in a healthy Caucasian population. *Lipids* 42:143–50.

Perez-Martinez, P., Perez-Jimenez, F., Ordovas, J.M., et al. (2007c). The APOB -516C/T polymorphism has no effect on lipid and apolipoprotein response following changes in dietary fat intake in a healthy population. *Nutr Metab Cardiovasc Dis* 17:224–29.

Rankinen, T., Church, T., Rice, T., et al. (2007). Effect of endothelin 1 genotype on blood pressure is dependent on physical activity or fitness levels. *Hypertension* 50:1120–5.

Robitaille, J., Houde, A., Lemieux, S., Gaudet, D., Perusse, L. and Vohl, M.C. (2007a). The lipoprotein/lipid profile is modulated by a gene-diet interaction effect between polymorphisms in the liver X receptor-alpha and dietary cholesterol intake in French-Canadians. *Br J Nutr* 97:11–18.

Robitaille, J., Houde, A., Lemieux, S., Perusse, L., Gaudet, D. and Vohl, M.C. (2007b). Variants within the muscle and liver isoforms of the carnitine palmitoyltransferase I (CPT1) gene interact with fat intake to modulate indices of obesity in French-Canadians. *J Mol Med* 85:129–37.

Saijo, Y., Yoshioka, E., Fukui, T., et al. (2007). Effects of the interaction between interleukin-6 -634C/G polymorphism and smoking on serum C-reactive protein concentrations. *Hypertens Res* 30:593–99.

Santos, J.L., Boutin, P., Verdich, C., et al. (2006). Genotype-by-nutrient interactions assessed in European obese women. A case-only study. *Eur J Nutr* 45:454–62.

Santosa, S., Demonty, I., Lichtenstein, A.H., Ordovas, J.M. and Jones, P.J. (2007). Single nucleotide polymorphisms in ABCG5 and ABCG8 are associated with changes in cholesterol metabolism during weight loss. *J Lipid Res* 48:2607–13.

Scacchi, R., Pinto, A., Rickards, O., et al. (2007). An analysis of peroxisome proliferator-activated receptor gamma (PPAR-gamma 2) Pro12Ala polymorphism distribution and prevalence of type 2 diabetes mellitus (T2DM) in world populations in relation to dietary habits. *Nutr Metab Cardiovasc Dis* 17:632–41.

Servin, B. and Stephens, M. (2007). Imputation-based analysis of association studies: candidate regions and quantitative traits. *PLoS Genet* 3:e114.

Shen, J., Arnett, D.K., Peacock, J.M., et al. (2007). Interleukin1beta genetic polymorphisms interact with polyunsaturated fatty acids to modulate risk of the metabolic syndrome. *J Nutr* 137:1846–51.

Smith, U. (2007). TCF7L2 and type 2 diabetes – we WNT to know. *Diabetologia* 50:5–7.

Sofi, F., Fatini, C., Sticchi, E., et al. (2007). Fish intake and LPA 93C>T polymorphism: gene-environment interaction in modulating lipoprotein (a) concentrations. *Atherosclerosis* 195:e147–e154.

Stephens, J.W., Dhamrait, S.S., Mani, A.R., et al. (2008). Interaction between the uncoupling protein 2-866G>A gene variant and cigarette smoking to increase oxidative stress in subjects with diabetes. *Nutr Metab Cardiovasc Dis* 18:7–14.

Subbiah, M.T. (2007). Nutrigenetics and nutraceuticals: the next wave riding on personalized medicine. *Transl Res* 149:55–61.

Talmud, P.J. (2007). Gene–environment interaction and its impact on coronary heart disease risk. *Nutr Metab Cardiovasc Dis* 17:148–52.

Tanaka, T., Delgado-Lista, J., Lopez-Miranda, J., et al. (2007a). Scavenger receptor class B type I (SCARB1) c.1119C>T polymorphism affects postprandial triglyceride metabolism in men. *J Nutr* 137:578–82.

Tanaka, T., Ordovas, J.M., Delgado-Lista, J., et al. (2007b). Peroxisome proliferator-activated receptor alpha polymorphisms and postprandial lipemia in healthy men. *J Lipid Res* 48:1402–8.

Teo, Y.Y. (2008). Common statistical issues in genome-wide association studies: a review on power, data QC, genotype calling and population structure. *Curr Opin Lipidol* 19:133–43.

Tracy, R. (2008). 'Deep phenotyping': characterizing populations in the era of genomics and systems biology. *Curr Opin Lipidol* 19:151–57.

Tsujita, Y., Nakamura, Y., Zhang, Q., et al. (2007). The association between high-density lipoprotein cholesterol level and cholesteryl ester transfer protein TaqIB gene polymorphism is influenced by alcohol drinking in a population-based sample. *Atherosclerosis* 191:199–205.

van Dam, R.M. and Hu, F.B. (2005). Coffee consumption and risk of type 2 diabetes: a systematic review. *J Am Med Assoc* 294:97–104.

van Dam, R.M., Willett, W.C., Manson, J.E. and Hu, F.B. (2006). Coffee, caffeine, and risk of type 2 diabetes: a prospective cohort study in younger and middle-aged US women. *Diabetes Care* 29:398–403.

Volcik, K., Ballantyne, C.M., Pownall, H.J., Sharrett, A.R. and Boerwinkle, E. (2007). Interaction effects of high-density lipoprotein metabolism gene variation and alcohol consumption on coronary heart disease risk: the atherosclerosis risk in communities study. *J Stud Alcohol Drugs* 68:485–92.

Wang, X.L. and Wang, J. (2005). Smoking–gene interaction and disease development: relevance to pancreatic cancer and atherosclerosis. *World J Surg* 29:344–53.

Weiss, E.P., Brandauer, J., Kulaputana, O., et al. (2007). FABP2 Ala54Thr genotype is associated with glucoregulatory function and lipid oxidation after a high-fat meal in sedentary nondiabetic men and women. *Am J Clin Nutr* 85:102–08.

Williamson, D.A., Allen, H.R., Martin, P.D., Alfonso, A.J., Gerald, B. and Hunt, A. (2003). Comparison of digital photography to weighed and visual estimation of portion sizes. *J Am Diet Assoc* 103:1139–45.

Zeggini, E. and McCarthy, M.I. (2007). TCF7L2: the biggest story in diabetes genetics since HLA?. *Diabetologia* 50:1–4.

Zhang, L., Mukherjee, B., Ghosh, M., Gruber, S. and Moreno, V. (2008). Accounting for error due to misclassification of exposures in case-control studies of gene-environment interaction. *Stat Med* 10;27(15):2756–83.

CHAPTER

2

Translating Nutrigenomics Research into Practice: The Example of Soy Protein

Elaine S. Krul and Peter J. Gillies

OUTLINE

Nutrition and Genomics
ISBN: 978-0-12-374125-7

SUMMARY

Nutrigenomics – or 'nutritional genomics' – is the study of all the genetic factors that influence the biological response to diet and the impact diet has on gene expression. A review of published dietary intervention studies on the effects of soy protein on serum cholesterol reveals that nutrition researchers have, in many cases, overlooked many of the basic tenets of nutrigenomics thereby propagating a collection of inconsistent reports that raise questions about the cholesterol-lowering activity of soy protein. Had these studies been conducted with well-characterized soy proteins in carefully defined study populations, it is likely that a more consistent picture and more interpretable results would have emerged. The situation with soy protein is not uncommon in nutrition research and underscores the need for a nutrigenomics checklist to enable clearer interpretation, replication and comparisons of nutrition experiments; such a checklist would also facilitate indexing and mining of related studies in electronic databases.

INTRODUCTION

Molecular science can be leveraged at almost every step along the food chain from the farmer to the consumer. Driven by basic research in plant, animal and human genomics, agricultural biotechnology companies can now modify plants to produce desired nutrient profiles and food companies, leveraging the knowledge of genomic medicine, can offer new and novel products designed to optimize nutrition and health (Figure 2.1).

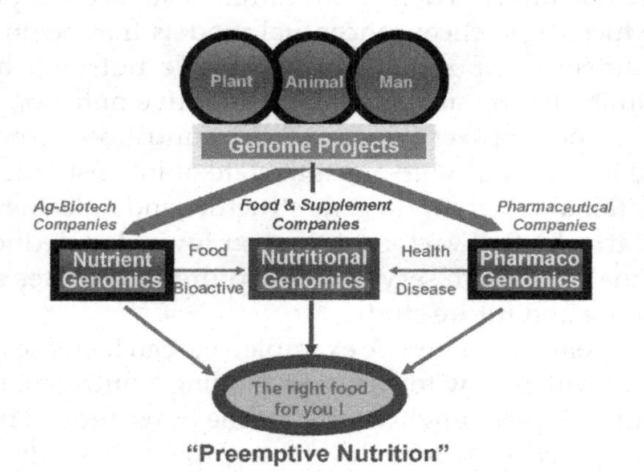

FIGURE 2.1 Basic genomic research in plants, animals and humans promises to offer novel products designed to deliver optimal nutrition and health in defined populations at various stages of life.

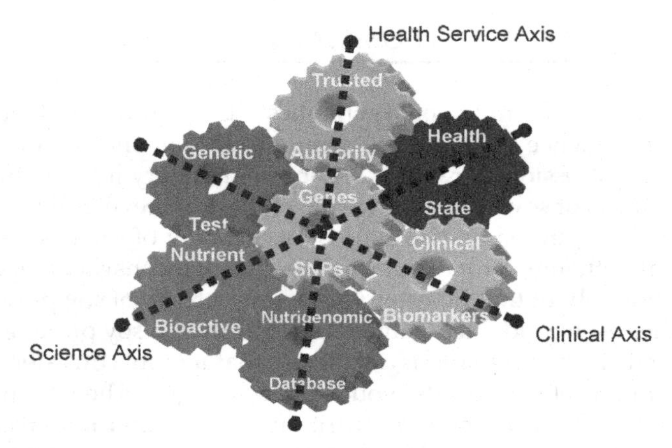

FIGURE 2.2 The pre-emptive nutrition model.

Indeed, at some point in the future, there is the exciting prospect of providing the 'right food for you' in the hopes of pre-emptively managing health and disease through diet.

Pre-emptive nutrition aspires to optimize health and offset the risk and/or progression of disease through an integrated understanding of genetic predisposition, nutrient–gene interaction and appropriately timed nutritional pharmacology (Gillies and Krul, 2007). The advent of a future state based on nutrigenomic principles depends upon the willingness and ability of science and society to build a health-centric supermarket based on sustainable business models in an enabling bioethical and legal environment (Castle et al., 2007).

Pre-emptive nutrition is a holistic multidimensional concept that is challenging to reduce to practice; conceptual models that begin to integrate the interdependent components of pre-emptive nutrition have interim operational utility. In one such model, pre-emptive nutrition is viewed as having three component axes (Figure 2.2). The nutrition science axis aligns nutrient–gene interactions with a health state of interest. The clinical axis factors in functional variations in gene structure and links these with clinical outcome variables that vector a consumer towards a deducible healthy phenotype. Finally, a health service axis captures, integrates and archives data for reference and future study.

Using soy proteins as a heuristic example, one can increase the granularity of the model with a view towards developing a nutrigenomic checklist that, if generally adopted, might accelerate the translation of nutrigenomic principles into practice. Before developing the case study, it is worthwhile reflecting on a new trend in the nutritional sciences, the advent of 'evidence-based nutrition'. Following the lead of the medical community, evidence-based nutrition (EBN) promises to provide a structured, unbiased analysis of the nature and strength of available evidence about the dietary intake and nutritional pharmacology of a nutrient (Balk et al., 2007). EBN is particularly valuable in benchmarking the state-of-the-science and identifying gaps in the evidence base that need to be filled through future research. EBN reports alert the research community to missing or needed data and provide guidance on how to avoid adding inadequate or repetitive data to the literature. Another use of EBN is to authenticate nutritional health claims (Center for Food Safety and Applied Nutrition (CFSAN), 2007). In this application, the current utility of the EB process is much less robust and generally leads to inconclusive and/or cautionary claims due to limited and discrepant clinical data. In this regard, it could be argued that nutrigenomic data may provide a stronger and more holistic underpinning to the EB review process. Recently, EBN reports by the Agency for Healthcare Research and Quality (AHRQ) and the American Heart Association (AHA) have called into question the health benefits of soy (Balk et al., 2005; Sacks et al., 2006). But what would these reports have looked like through the lens of nutrigenomics where defining the response haplotype or phenotype rather than drawing a 'one size fits all' efficacy statement was the goal? In this chapter, we explore this question using soy proteins as a case study.

THE NUTRITION SCIENCE AXIS

Soy Protein as a Cholesterol-Lowering Nutrient

The soy protein health claim was issued by the Food and Drug Administration (FDA) in 1999 and states that: 'Twenty-five grams of soy protein a day, as part of a diet low in saturated fat and cholesterol, may reduce the risk of heart disease'. In 2007, the FDA announced its intent to re-evaluate the scientific evidence for soy protein and the risk of coronary heart disease using its newly proposed 'evidence-based' guidelines. In 2005, the NIH commissioned a similar review of the health benefits of soy protein using an evidence-based approach (Balk et al., 2005); this AHRQ review serves as the frame of reference for the nutrigenomic discussion to follow.

The evidence-based review by the AHRQ concluded that, for the soy products used in the evidence-based analysis, the consumption of these products yielded a significant net reduction of low density lipoprotein (LDL) cholesterol (−5 mg/dl, 95% confidence interval [CI] −8 to −3 mg/dl) (Balk et al., 2005). There was, however, considerable variability in the cholesterol-lowering response. Using the nutrigenomic model of Figure 2.2, a key question to ask relates to the nature of the soy protein used in the various studies. As it turns out, many different sources of soy protein were used in these studies and they were consumed in a variety of

TABLE 2.1 Nutrient bioactive information provided in the AHRQ evidence-based review of soy protein studies of cholesterol reduction

	% of studies
Source of soy protein (brand name or manufacturer)	86
Protein isoflavone content analysis	49
Proximate analysis (ash, fat, carbohydrate and/or mineral content)	24
Protein processing information	8
Protein biochemical characterization (SDS-PAGE, SEC, etc.)	0

food forms including dry protein powder mixed into beverages and other foods, ready to drink beverages, meat analogs, tofu, yogurt and baked goods. Table 2.1 summarizes the information regarding the nature of the soy protein used in the 51 studies that served as the foundation of the report. Eighty six percent of the studies identified the source or manufacturer of the soy protein, however, this information offers little about the actual nature of the soy protein studied. About half the studies reported the isoflavone content of the soy protein, but it is not clear whether this reflects isoflavones isolated with the soy protein or isoflavones added back to the protein as extracts. One-quarter of the studies reported on the proximate analysis of the soy protein products used which could have revealed potential sources or confounders of any bioactivity that may be observed. Most remarkably, there was only minimal information provided on protein processing (8% of the studies mentioned that alcohol was used to deplete the protein of isoflavones) and no information provided on the biochemical or biophysical characterization of the protein. In essence, the clinical test article used across the 51 studies was an uncharacterized protein. While this is a general issue with nutrition and natural product research, it is an even larger problem in a nutrigenomic era as the structure–function relationship may equate to nutrient–gene interactions that are measured by exquisitely sensitive technologies.

Soy Proteins as Products by Process

For many reasons, dietary proteins are particularly challenging to incorporate into the nutrigenomic model. While proteins may be readily defined in terms of primary amino acid content, their rich functionality and nutritional pharmacology often resides in their higher order structure, component complexity and absorbable digestion products. Indeed, dietary proteins are rather unique among the macronutrients as they are often consumed in their 'non-native' form, particularly in Westernized societies. More specifically, dietary proteins are readily denatured (a change in structure without breaking covalent bonds) by a variety of processes including freezing, heat, precipitation, drying and whipping to name just a few. In

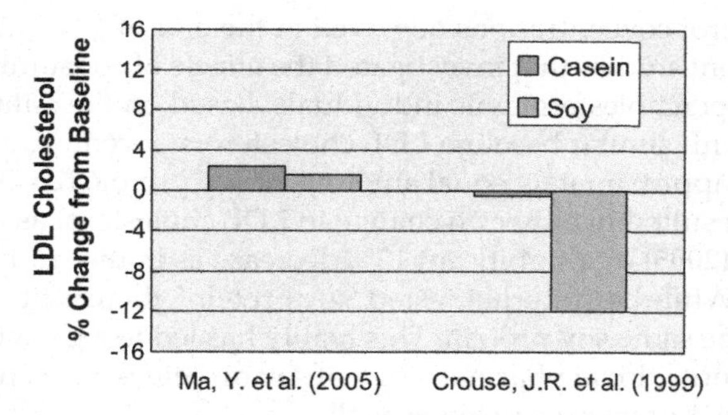

FIGURE 2.3 LDL cholesterol lowering observed in two separate studies using soy protein beverage supplementation as a dietary intervention. Study populations were similar and baseline LDL cholesterol levels averaged 170 ± 4 and 184 ± 4 mg/dl in the study groups from the **Ma et al. (2005)** and Crouse et al. (1999) studies, respectively. Study subjects consumed 31.5 g soy protein and 120 mg isoflavones per day for 5 weeks or 25 g soy protein and 62 mg isoflavones per day for 9 weeks as beverage supplements in the **Ma et al. (2005)** or Crouse et al. (1999) studies, respectively. The soy protein preparations used in the two studies were processed by different methods.

addition, proteins are subjected to non-enzymatic amino acid modification during processing or storage (glycosylation and Maillard reactions and oxidation), which can lead to losses of nutritive amino acids, generation of dehydro or cross-linked amino acids, glycosylamines or diamino sugars or changes in the ability of the protein to be digested by intestinal proteases. In addition, dietary protein can be consumed in minimally to wholly enzymatically hydrolyzed forms (as a result of fermentation and/or enzyme addition, e.g. yogurts, cheeses, infant formulas, some soy protein isolates, miso, soy sauce and energy drinks). Additionally, proteins in the diet also carry many minerals and water-soluble vitamins and phytochemicals as non-covalently attached components that are retained or removed to various and variable degrees during processing and/or preparation of the protein for consumption. Notably, processing is primarily directed at achieving some desired food formulation functionality as opposed to enhancing some aspect of a protein's nutritional pharmacology.

Soy protein is no exception to this world of proteinacious complexity with novel soy products being consumed in an ever-increasing variety of forms that have been differentially processed in some way to optimize food functionality. While generally retaining its nutritive value as a high quality protein in these food forms, some researchers have pointed out that differences in the physicochemical composition of the soy product may largely account for differences in clinical and epidemiological outcomes of soy protein studies (Gianazza et al., 2003; Erdman et al., 2004). As a case in point, Figure 2.3 illustrates the disparate LDL cholesterol-lowering activity of two soy protein beverages prepared with differently processed soy protein preparations. The data show the percent changes in plasma

LDL cholesterol concentrations observed at the end of a treatment period in two different studies that investigated the effects of consuming soy beverages by hypercholesterolemic individuals. Based on the data of subjects presenting with similar baseline LDL cholesterol concentrations, the consumption of approximately equal amounts of soy protein (25–32 g/day for 5–9 weeks) resulted in either no change in LDL cholesterol as in the study by **Ma et al. (2005)** or a significant 12% decrease as in the study by Crouse et al. (1999). While both studies tested 'soy protein', there is little assurance they tested the same soy protein. This reality has led to a growing recognition that the nutritional pharmacology of soy protein is a function not only of the protein, but its processing as well.

Gianazza et al. (2003) suggested differences in the ability of soy protein preparations to lower cholesterol were related to the degree of protein hydrolysis in the various preparations. For example, Cholsoy®, the soy protein preparation used in many European studies, results in 20% or greater reductions in LDL cholesterol (**Anderson, 1995**), whereas Supro® soy, the soy isolate commonly used in many North American studies (Baum et al., 1998; Crouse et al., 1999), tends to elicit much smaller decreases in LDL cholesterol. Compared to Supro®, Cholsoy® is a relatively unrefined protein preparation, containing approximately 30% carbohydrate by weight (oligosaccharides and non-starch polysaccharides). Not unexpectedly, Cholsoy® and Supro® proteins exhibit significantly different patterns of molecular weight species by two-dimensional electrophoresis (Gianazza et al., 2003). A recent prospective study assessing the efficacy of soybean products with minimal to no enzymatic processing in mildly hypercholesterolemic subjects showed only a very modest lowering of LDL cholesterol, relative to animal protein, when consumed at a level of approximately 37.5 g in the context of the National Cholesterol Education Program (NCEP) Step II diet (Matthan et al., 2007). Notably, the soymilk diet (which included soymilk, tofu and soy yogurt) was the only group that showed a significant reduction in LDL cholesterol relative to the animal protein diet. In a recent retrospective analysis of 406 middle-aged women and men in China, a dose-dependent decrease in total and LDL cholesterol was associated with increasing intake of soy foods commonly found in Asia, e.g. various forms of tofu, tofu pudding/soymilk, soybean sprouts, soybeans and fermented soy products (Zhang et al., 2007). The median intakes in this study ranged from 0.72 and 5.73 g soy protein/day for men and from 0.89 to 5.89 g soy protein/day for women. It is worth mentioning that, at these consumption levels, a significant dose-dependent decrease in carotid intima-media (IMT) thickness was detected, suggesting a vascular benefit of dietary soy protein in these food forms (Zhang et al., 2007).

As mentioned earlier, soy protein is a complex macronutrient and other components commonly found in soy protein preparations may impact its cholesterol-lowering activity. Notable among such components are isoflavones, phospholipids and carbohydrates. There has been considerable debate about whether the isoflavones present in soy protein contribute to its cholesterol lowering properties. Human studies using purified isoflavone supplements have consistently failed to show significant effects on serum lipid concentrations (Dewell et al., 2006) and the early studies on soy protein showed significant cholesterol lowering with preparations essentially devoid of isoflavones (Sirtori et al., 2007). Alcohol-washed soy protein, which is depleted of isoflavones, has been used in many studies as the 'control' protein against which soy protein with isoflavones was compared. The process of alcohol-washing also removes other alcohol-extractable phytochemicals, such as saponins, phytic acid and, possibly, peptides generated during enzymatic processing (Gianazza et al., 2003), however, it may also alter the soy protein's physicochemical properties, all of which confounds interpretation of these studies. Just as the debate concerning isoflavones appeared to be resolved, a recent study once again raised the question as to whether isoflavones may indeed play a role in cholesterol reduction. Clerici et al. (2007) reported that the consumption of a once daily soy germ enriched pasta containing 33 mg isoflavones as aglycones (and negligible soy protein, 0.8 g/serving) elicited a statistically significant 8.6% reduction in LDL cholesterol in adult men and women after 4 weeks compared to conventional pasta. It is important to note that 69% of the subjects in the Clerici study were equol producers, i.e. their gut microflora efficiently converted daidzein to equol, and this group exhibited the significant reductions in total and LDL cholesterol (Clerici et al., 2007). This novel observation raises the possibility of a metabolic phenotype that contributes to the soy protein response profile. Similar to the isoflavone story, there is a body of literature related to the impact of phospholipid content on soy protein's ability to lower serum cholesterol. Many studies in Japan have used hydrolyzed soy proteins that repeatedly show cholesterol-lowering benefits at 6–30 g/day (Wang et al., 1995; Imura et al., 1996; Hori et al., 2001). The presence of phospholipids added during processing (lecithination) (Hori et al., 2001) appears to enhance the cholesterol-lowering effect of the soy hydrolysates. In this regard, it should be noted that many soy protein preparations used in human studies contain lecithin to varying degrees. In summary, protein processing and its effects on adventitious components clearly impact the cholesterol-lowering activity of soy protein products.

Processing and Biochemical Characterization of Soy Proteins

Figure 2.4 is a schematic representation of the soy protein processing procedure. Soybeans are a complex plant material. During protein processing, most of the non-protein molecular species are separated from the protein.

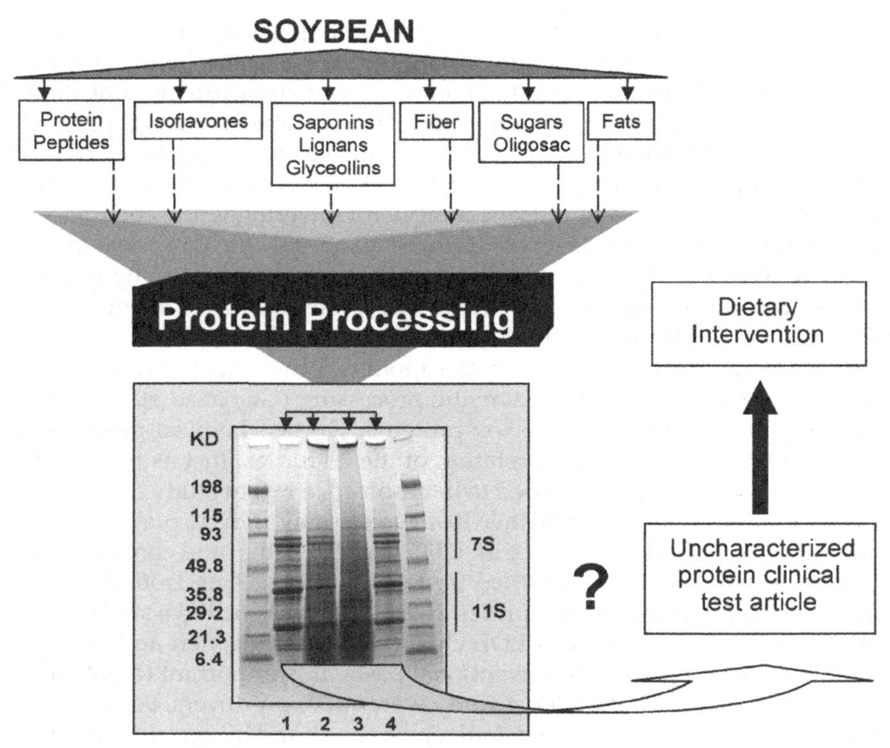

FIGURE 2.4 Schematic representation of the soy processing procedure. The inset depicts an SDS-PAGE of four different soy protein preparations made by four distinct processing methods. Dietary intervention studies using soy protein may have used a variety of soy protein preparations like those depicted in the electrophoretic gels in the inset, however, most dietary intervention studies have not described any characterization of the proteins used.

As the protein undergoes isolation and processing, it undergoes structural, compositional and chemical changes; unfortunately, these changes are largely undocumented in the 'Materials' section of scientific publications. The proteins depicted in the electrophoretic gel in Figure 2.4 are representative of protein products developed for optimal performance in specific food applications. Contrast the protein products run in lanes 1 and 4 used for neutral beverage and emulsified meat applications, respectively, with the proteins in lanes 2 and 3; the latter are produced using enzymatic digestion of the proteins, a method often used for making proteins more soluble in neutral as well as acid or clear beverage applications. Different enzymatic processes yield soy protein products with different degrees of hydrolysis and peptide lengths, with the protein in lane 3 undergoing more hydrolysis than the protein in lane 2. It is important to mention that while SDS-PAGE (a protein separating technique) provides information on the molecular weight distributions of a product, there can still be wide functionality differences between proteins that share a common SDS-PAGE profile. In this regard, SDS-PAGE is but one of several analytical methods needed to characterize proteins.

Clearly, there is an urgent need for research into the impact of protein processing on the digestibility, bioavailability and health effects of dietary proteins. The success of this research depends upon the mutual cooperation of industry, academia and journal editors. Given that the food industry is the principal supplier of test proteins to the nutrition science community, it is critical that they recognize the importance of documenting the basic characterization of the protein used in research and commerce. Similarly, researchers need to harmonize their study protocols to enable cross-study comparisons, and journal editors, at least in an ideal world, could require both of the above as a condition of publication. Lemay et al. (2007) have recently proposed a blueprint of how nutrition data collected with standardized protocols and standardized reporting procedures enables the techniques of bioinformatics to be applied in nutrition research. Such standardization and bioinformatics-enabled analyses would revolutionize the evidence-based review process and streamline the review system for the scientific evaluation of health claims.

THE PROPOSED MINIMUM INFORMATION ABOUT A DIETARY PROTEIN (MIADP) STANDARD

Standardization of genomic and proteomic data has already become an accepted practice in many fields of research and is increasingly required by journals as a prerequisite for manuscript review. For example MIAME (Minimal Information About a Microarray Experiment) has been developed to facilitate the reporting and indexing of the vast and complex data sets of transcriptome experiments (Brazma et al., 2001). The MIAME standard is now required for publication of microarray experiments in many key journals. The journals also require that authors provide accession or identifier numbers to databases that identify specific molecular structures e.g. the protein sequence database, UniProt (The UniProt Consortium, 2007) and protein structure database, Protein Data Bank (PDB) (Berman et al., 2000). It is not unreasonable, therefore, to propose the development of a dietary protein database that contains all the key data about dietary proteins that have been used in investigational studies. To this end, we propose that a standard for dietary protein data be established, i.e. a Minimum Information About a Dietary Protein (MIADP). In keeping with the intent and spirit of the MIAME standard, every effort should be made to find a middle ground between overburdening researchers and journals or

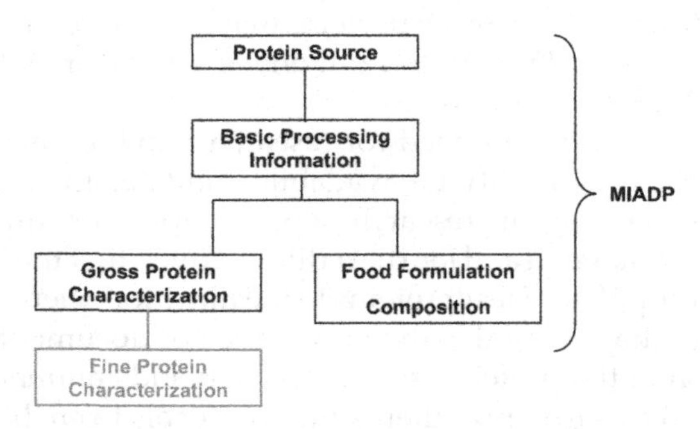

FIGURE 2.5 Schematic representation of the major components for a proposed standard to provide the minimum information about a dietary protein (MIADP) for all dietary intervention trials evaluating proteins.

comprising the proprietary information of companies while still reporting key data in sufficient detail to be useful and in a format that allows migration into electronic databases.

Based on the principles of the MIAME standard we propose the following MIADP standard as a starting point for dietary proteins used in research and commerce. The basics of the MIADP standard are illustrated in Figure 2.5 and include the following elements:

1. Protein source: the plant, animal or fungal species and strain which is the source of the protein should be reported. Where appropriate, information should be given as to whether the source of protein is a genetically modified or specifically bred, natural strain of the species.
2. Basic processing information: information about the processing methods used to prepare the protein product should be provided in sufficient detail to enable an appreciation of the final state of the protein in the product. Specific information to be provided could include: a) protein extraction method (starting material, organic/aqueous extraction, membrane and/or chemical separation methods, etc.); b) enzymatic treatments or degree of hydrolysis; c) additives during processing (e.g. lecithin, calcium); and d) drying methods should be described (e.g. spray drying).
3. Gross protein characterization: the protein product must be characterized by clearly defined attributes that allow other researchers to compare the different protein products used in dietary interventions. We propose that the following attributes be reported: a) physical characterization (e.g. particle size, protein surface hydrophobicity (8-anilino-1-naphthalene-sulfonate (ANS) fluorescence), surface charge

(zeta potential), percent degree of hydrolysis, etc.); b) proximate analysis (lipid, carbohydrate, ash and moisture content); c) amino acid composition (including modified amino acids); d) molecular weight profiling by SDS-PAGE and/or size exclusion chromatography (SEC); e) quantitative analysis of phytochemicals likely to be associated with the protein of interest (e.g. isoflavones, saponins, phytate and sterols); and f) measurements of functionality such as solubility and viscosity.

4. Food formulation composition: protein ingredients can be delivered in a variety of food forms and matrices. The same ingredient delivered in different food forms may be subject to additional protein physical and chemical modifications and may exhibit different digestibility and bioavailability in the intestine when combined in different matrices once consumed. Therefore, information about the preparation and detailed nutrient composition of the final food form/matrix should be provided.
5. Fine protein characterization: in cases where a refined and/or purified protein (e.g. soy 7S protein) is used in dietary intervention studies, additional information about the protein product may be warranted such as protein sequence and higher order structural information.

CHARACTERIZED SOY PROTEINS ENABLES MECHANISM OF ACTION RESEARCH

One of the major benefits of working with well-defined and characterized soy proteins is likely to be the elucidation of a credible and testable mechanism of action of the protein's cholesterol-lowering activity. One mechanism that could account for the cholesterol-lowering properties of hydrolyzed soy protein is bile acid binding. Support for this mechanism comes from studies showing a two–threefold increase in fecal bile acid excretion in young women consuming 4% of their total energy as an undigested fraction of soy hydrolysate (Wang et al., 1995). This mechanism is further supported by animal studies showing increases in fecal bile acid excretion in rats consuming soy protein hydrolysates (Iwami et al., 1986; Sugano et al., 1988; Ogawa et al., 1992; Higaki et al., 2006). Of note, the increases in fecal bile acid excretion were dose-dependent (Sugano et al., 1988) and observed whether the rats were maintained on normal or cholesterol-supplemented diets (Sugano et al., 1988; Ogawa et al., 1992; Higaki et al., 2006). Iwami et al. (1986) proposed that the hydrophobicity of plant-derived proteins that were relatively resistant to digestion in the gut accounted for the sequestration of bile acids. More recently, specific peptide sequences derived from soy protein have been identified that have bile acid binding activity (**Choi et al., 2002**; Pak et al., 2005). A related gut-based mechanism relates to the ability of soy protein to disrupt cholesterol absorption. Investigators have identified soy hydrolysates with the ability to disrupt cholesterol micellarization *in vitro* (**Nagaoka et al., 1999**; Zhong et al., 2007); these soy hydrolysates or isolated fractions also exhibited cholesterol-lowering effect in rats (**Nagaoka et al., 1999**; Zhong et al., 2007). Whether the bile acid binding and micellar disruptive activities of these peptide mixtures result from the same molecular mechanisms or some other distinct mechanisms remains to be determined (**Nagaoka et al., 1999**).

An important variation of the peptide-based mechanism involves the potential for direct effects of specific soluble peptides acting on the liver that have intrinsic cholesterol-lowering activity (Lovati et al., 2000). While there has been no direct evidence of this *in vivo*, the *in vitro* data coupled with what is now known about the effects of processing on soy proteins has interesting implications. For example, specific peptide sequences derived from the soy 7S globulin have been shown to upregulate LDL receptor activity in HepG2 cells (Manzoni et al., 1998; Lovati et al., 2000). A separate group recently demonstrated stimulation of LDL receptor gene transcription in HepT9A4 cells by soy peptide hydrolysate fractions when presented at similar peptide concentrations used in the studies described above (Cho et al., 2007), but the specific peptides responsible for this bioactivity were not identified. Yoshikawa et al. (2000) reported that administration of 50 mg/kg of a peptide derived from soy glycinin (Leu-Pro-Tyr-Pro-Arg) to 4-week-old ICR mice for 2 days resulted in a 25% reduction in serum cholesterol without any apparent change in fecal cholesterol or bile acids. Rho et al. (2007) demonstrated that a low molecular weight fraction of black soy peptide hydrolysate (<10 KDa) fed to young rats (10% of their energy as soy peptide and 10% as casein) on a high fat, high cholesterol diet resulted in a 25% lower serum cholesterol level than a casein control group (20% of energy as casein).

In conclusion, animal and human studies support the concept that soy proteins have an inherent ability to lower serum cholesterol through well established and clinically validated mechanisms of action. This sets the stage to focus the nutrigenomic lens on these cholesterol-lowering mechanisms to explore underlying nutrient–gene interactions (Kleemann et al., 2007) and the impact of genetic variations.

INTERSECTION OF THE NUTRITION SCIENCE AND CLINICAL AXES: DEFINING THE 'RESPONSE PHENOTYPE'

Only three of the 51 studies summarized in the AHRQ review reported any attempt to genotype the subjects for any polymorphisms known to be involved in cholesterol metabolism (these three studies identified subjects with familial hypercholesterolemia). Two of the 51 studies investigated soy protein efficacy in type II diabetic subjects. The criteria for study subject selection in the rest of the 51 studies were fairly broad and included men and pre- and post-menopausal women of various ages and body mass index. Baseline clinical measurements were made in most, but not all, cases after a defined dietary run-in period. The run-in and intervention diets were not standardized across the studies and varied significantly (26 studies used habitual diets, 13 studies specified diets and nine, two and one study used the NCEP Step I, Step II and Therapeutic Lifestyle Changes (TLC) diets, respectively). In effect, the AHRQ endeavored to draw general conclusions on the ability of soy protein to lower serum cholesterol as a function of (variable soy protein) × (variable background diet) × (variable subject genetics). Even a potent cholesterol-lowering drug would be challenged in such a scenario and it should not be surprising that the cholesterol-lowering activity of soy protein regresses to some low mean effect with a small signal to noise ratio. When well defined and carefully characterized soy proteins (or peptides) finally become available, the research community will be in a unique position to ask the nutrigenomic question: what is the response phenotype for soy protein's effect on serum cholesterol? In the interim, the molecular pharmacology literature is quite instructive in terms of candidate genes and single nucleotide poymorphisms (SNPs) that may be involved in defining the response phenotype.

Polymorphisms in regulatory genes in the cholesterol synthesis pathway (e.g. 3-hydroxy-3-methylglutaryl coenzyme A reductase (HMG-CoA reductase) or absorption and transport pathways (e.g. Niemann-Pick C1-like protein 1 precursor (NPC1L1)) have been shown to affect an individual's response to different cholesterol-lowering drugs such as pravastatin and ezetimibe (**Chasman et al., 2004**; Simon et al., 2005). In addition, a polymorphism in the sterol receptor element binding protein 1c (SREBP-1c gene) has recently been found to be associated with an increased cholesterol-absorption inhibition in response to ezetimibe in humans (Berthold et al., 2007). Polymorphisms in other genes involved in cholesterol absorption, namely, the half transporters ABCG5 and ABCG8, have also been shown to affect cholesterol absorption and synthesis in women during weight loss (Santosa et al., 2007). The molecular precedents have clearly been set; thus, it would not be surprising if individuals with these types of genetic variations also exhibited differential responses to soy protein as well.

The cholesterol-lowering response to soy protein has been shown to be influenced by apolipoprotein E genotype in familial hypercholesterolemic (FH) subjects; specifically, E3/E3 and E3/E4 subjects experience significantly larger decreases in total cholesterol than E3/E2 subjects (Gaddi et al., 1991). This observation is consistent with an ability of soy protein to depress the low level of hepatic LDL receptor activity in individuals having the more efficient LDL receptor binding E3 and E4 alleles. However, not all studies that have segregated cholesterol-lowering data on the basis of apoE alleles have shown this difference for soy protein (**Vigna et al., 2000**). This could indicate that the LDL receptor gene defect in the FH subjects followed in the Gaddi study could involve an epistatic effect on the apoE genotype. These data also indicate that the apoE genotype could be a 'response modifier' rather than a direct participant in the mechanism whereby soy protein affects cholesterol homeostasis. The literature on the genetics of response modifying apolipoproteins and their impact on dietary interventions is rich and well documented (Ordovas, 2006).

As discussed earlier in this chapter, isolated soy isoflavones do not appear to exert a significant LDL cholesterol-lowering effect except in subjects who metabolize daidzein to equol (Clerici et al., 2007). In a *post-hoc* analysis of data from a soy dietary intervention study, Meyer et al. (2004) noted that the cholesterol 'responders' in their soy intervention study were also equol producers. The equol producing phenotype appears to be under genetic control, and the specific intestinal bacteria responsible for the dehydroxylation and reduction of daidzein to equol have not been identified (Frankenfeld et al., 2004). Setchell et al. (2002) proposed 'bacterio-typing' people according to their ability to produce equol prior to randomization into dietary intervention studies that use isoflavone containing ingredients, since the production of equol may well be a confounding variable in such studies. Another response modifier in the equol story may be polymorphisms of estrogen receptor β (ERβ) recognizing that equol is an estrogenic compound. For instance, pre-menopausal and post-menopausal women on exogenous estrogen carrying the (G/A) genotype of the ERβ 1082G>A polymorphism have significantly lower total and LDL cholesterol than their counterparts with the (G/G) genotype (**Almeida et al., 2005**) and post-menopausal women bearing the A/A genotype of the ERβ splice variant polymorphism (ERβ (cx) Tsp509I) show significantly higher HDL cholesterol levels after isoflavone consumption than their G/A or G/G counterparts (Hall et al., 2006). Thus, it is important to consider the microbiome in host lipid metabolism. This is underscored in a recent report from Martin et al. (2007) who demonstrated that the specific gut microbial populations in mice significantly influenced bile acid metabolism and activity and that substitution of a conventional mouse microbiome with a human infant microbiome altered the bile acid metabolism such that there was a significant reduction in plasma lipids and increased hepatic lipid concentrations.

In conclusion, genetic factors are clearly involved in determining the 'response phenotype' to soy protein in clinical studies. Future studies using defined soy proteins would benefit greatly from using genetically characterized study populations in standardized interventional trials

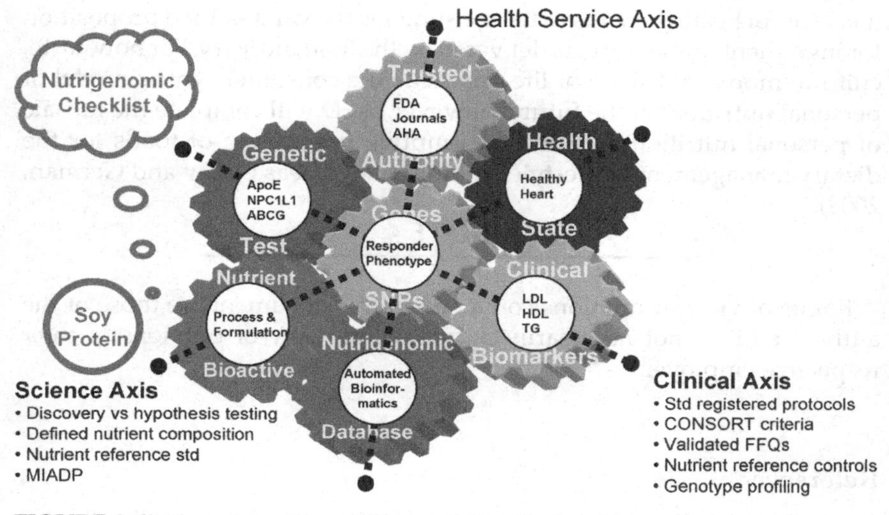

FIGURE 2.6 The pre-emptive nutrition model in practice with soy protein.

(Figure 2.6). In addition, care should be taken to identify environmental factors that can influence the 'response phenotype' through gene–environment interactions (such as smoking, alcohol consumption and physical activity). The latter is described in detail in Chapter 1.

CONCLUSION

Nutrigenomics is a powerful tool that guides investigators though a more global and molecular consideration of the various factors that influence the human biological response to diet. Using harmonized protocols for nutritional intervention studies with a nutrigenomic focus helps identify a broader spectrum of critical study parameters and reduces the background noise of confounding variables and factors. The nutrigenomic 'lens' allows one to view a complex data-rich landscape and effectively zoom in on the biologic effect elicited by the bioactive nutrient of interest. The judicious and harmonious application of nutrigenomic principles to dietary intervention trials could increase the signal to noise of the relevant nutritional pharmacology and enable researchers to confidently detect subtle but meaningful biologic effects of dietary components. This depth of science comes at a cost to the business side of nutrition, however, as molecular nutrition will inevitably erode the mass marketing paradigm of 'one size fits all' and lead to consumer segmentation. In this regard, nutrigenomic segmentation simply pushes the limits of 'personal nutrition' into a space that business models have yet to exploit, but need to in the future, if truly meaningful health benefits to the consumer is the value-added proposition for ingredients and products delivered by the food industry. For now, taste, cultural mores and stage of life dominate the consumer-centric model of personal nutrition; in the future, however, R&D will complete the mosaic of personal nutrition with the pre-emptive positioning of foods for the dietary management of lifestyle and inherited diseases (Fay and German, 2008).

NOTE

Points of view or opinions contained in this document are those of the authors and do not necessarily represent the view or opinions of their respective employer.

References

Balk, E., Chung, M., Chew, P. et al. (2005). Effects of soy on health outcomes. AHRQ Publication No. 05-E024-2. Rockville, MD: Agency for Healthcare Research and Quality.

Balk, E.M., Horsley, T.A. and Newberry, S.J. (2007). A collaborative effort to apply the evidence-based review process to the field of nutrition: challenges benefits, and lessons learned. *Am J Clin Nutr* 85:1448–56.

Baum, J.A., Teng, H., Erdman, J.W. Jr., et al. (1998). Long-term intake of soy protein improves blood lipid profiles and increases mononuclear cell low-density-lipoprotein receptor messenger RNA in hypercholesterolemic, postmenopausal women. *Am J Clin Nutr* 68: 545–51.

Berman, H.M., Westbrook, J., Feng, Z., et al. (2000). The protein data bank. *Nucleic Acids Res* 28:235–42.

Berthold, H.K., Laaksonen, R., Lehtimaki, T., et al. (2007). SREBP-1c gene polymorphism is associated with increased inhibition of cholesterol-absorption in response to ezetimibe treatment. *Exp Clin Endocrinol Diab* 116:262–7.

Brazma, A., Hingamp, P., Quackenbush, J., et al. (2001). Minimum information about a microarray experiment (MIAME) – towards standards for microarray data. *Nat Genet* 29:365–71.

Castle, D., Cline, C., Daar, A.S., Tsamis, C. and Singer, P.A. (2007). *Science, society, and the supermarket: opportunities and challenges of nutrigenomics*. John Wiley & Sons, Inc.

CFSAN/Office of Nutrition, Labeling and Dietary Supplements (2007). Guidance for industry: evidence-based review system for the scientific evaluation of health claims. http://www.cfsan.fda.gov/~dms/hclmgui5.html.

Chasman, D.I., Posada, D., Subrahmanyan, L., et al. (2004). Pharmacogenetic study of statin therapy and cholesterol reduction. *JAMA* 291:2821–7.

Cho, S., Juillerat, M.A. and Lee, C. (2007). Cholesterol lowering mechanism of soybean protein hydrolysate. *J Ag Food Chem* 55:10599–604.

Clerici, C., Setchell, K.D.R., Battezzati, P.M., et al. (2007). Pasta naturally enriched with isoflavone aglycones from soy germ reduces serum lipids and improves markers of cardiovascular risk. *J Nutr* 137:2270–8.

Crouse, J.R., Morgan, T., Terry, J.G., Ellis, J., Vitolins, M. and Burke, G. (1999). A randomized trial comparing the effect of casein with that of soy protein containing varying amounts of isoflavones on plasma concentrations of lipids and lipoproteins. *Arch Intern Med* 159: 2070–6.

Dewell, A., Hollenbeck, P.L.W. and Hollenbeck, C.B. (2006). Clinical review: a critical evaluation of the role of soy protein and isoflavone supplementation in the control of plasma cholesterol concentrations. *J Clin Endocrinol Metab* 91:772–80.

Erdman, J.W. Jr., Badger, T.M., Lampe, J.W., Setchell, K.D.R. and Messina, M. (2004). Not all soy products are created equal: caution needed in interpretation of research results. *J Nutr* 134:1229S–33.

Fay, L. and German, B.A. (2008). Personalizing foods: is genotype necessary? *Curr Opin Biotechnol* 19:121–28

Frankenfeld, C.L., Atkinson, C., Thomas, W.K., et al. (2004). Familial correlations, segregation analysis, and nongenetic correlates of soy-isoflavone-metabolizing phenotypes. *Exp Biol Med* 229:902–13.

Gaddi, A., Ciarrocchi, A., Matteucci, A., et al. (1991). Dietary treatment for familial hypercholesterolemia – differential effects of dietary soy protein according to the apolipoprotein E phenotypes. *Am J Clin Nutr* 53:1191–6.

Gianazza, E., Eberini, I., Arnoldi, A., Wait, R. and Sirtori, C. (2003). A proteomic investigation of isolated soy proteins with variable effects in experimental and clinical studies. *J Nutr* 133:9–14.

Gillies, P.J. and Krul, E.S. (2007). Using genetic variation to optimize nutritional preemption. *J Nutr* 137:270S–74.

Hall, W.L., Vafeiadou, K., Hallund, J., et al. (2006). Soy-isoflavone enriched foods and markers of lipid and glucose metabolism in postmenopausal women: interactions with genotype and equol production. *Am J Clin Nutr* 83:592–600.

Higaki, N., Sato, K., Suda, H., et al. (2006). Evidence for the existence of a soybean resistant protein that captures bile acid and stimulates its fecal excretion. *Biosci Biotechnol Biochem* 70:2844–52.

Hori, G., Wang, M., Chan, Y., et al. (2001). Soy protein hydrolysate with bound phospholipids reduces serum cholesterol levels in hypercholesterolemic adult male volunteers. *Biosci Biotechnol Biochem* 65:72–8.

Imura, T., Tanaka, M., Watanabe, T., Kudo, S., Uchida, T. and Kanazawa, T. (1996). Serum lipid improvement effect of soy protein: Study on minimum effective intake of soy protein in humans. *Therap Res* 17:1–12.

Iwami, K., Sakakibara, K. and Ibuki, F. (1986). Involvement of post-digestion 'hydrophobic' peptides in plasma cholesterol-lowering effect of dietary plant proteins. *Agric Biol Chem* 50:1217–22.

Kleemann, R., Verschuren, L., van Erk, M.J. et al. (2007). Atherosclerosis and liver inflammation induced by increased dietary cholesterol intake: a combined transcriptomics and metabolomics analysis. *Genome Biol* 8:R200 (1–15).

Lemay, D.G., Zivkovic, A.M. and German, J.B. (2007). Building the bridges to bioinformatics in nutrition research. *Am J Clin Nutr* 86:1261–9.

Lovati, M.R., Manzoni, C., Gianazza, E., et al. (2000). Soy protein peptides regulate cholesterol homeostasis in HepG2 cells. *J Nutr* 130:2543–9.

Ma, Y., Chiriboga, D., Olendzki, B.C., et al. (2005). Effect of Soy Protein containing isoflavones on blood lipids in moderately hypercholesterolemic adults: A randomized controlled trial. *J Am Coll Nutr* 24:275–85.

Manzoni, C., Lovati, M.R., Gianazza, E., Morita, Y. and Sirtori, C. (1998). Soybean protein products as regulators of liver low-density lipoprotein receptors. II. A-a′ rich commercial soy concentrate and a′ deficient mutant differently affect low-density lipoprotein receptor activation. *J Ag Food Chem* 46:2481–4.

Martin, F-P.J., Dumas, M.-E., Wang, Y. et al. (2007). A top-down systems biology view of microbiomemammalian metabolic interactions in a mouse model. *Mol Systems Biol* 3:112, doi: 10.1038/msb4100153

Matthan, N.R., Jalbert, S.M., Ausman, L.M., Kuvin, J.T., Karas, R.H. and Lichtenstein, A.H. (2007). Effect of soy protein from differently processed products on cardiovascular disease risk factors and vascular endothelial function in hypercholesterolemic subjects. *Am J Clin Nutr* 85:960–66.

Meyer, B.J., Larkin, T.A., Owen, A.J., Astheimer, L.B., Tapsell, L.C. and Howe, P.R.C. (2004). Limited lipid-lowering effects of regular consumption of whole soybean foods. *Ann Nutr Metab* 48:67–78.

Ogawa, T., Gatchalian-Yee, M., Sugano, M., Kimoto, M., Matsuo, T. and Hashimoto, Y. (1992). Hypocholesterolemic effect of undigested fraction of soybean protein in rats fed no cholesterol. *Biosci Biotechnol Biochem* 56:845–48.

Ordovas, J.M. (2006). Nutrigenetics, plasma lipids and cardiovascular risk. *J Am Dietetic Assoc* 106:1074–81.

Pak, V.V., Koo, M.S., Kasymova, T.D. and Kwon, D.Y. (2005). Isolation and identification of peptides from soy 11S-globulin with hypocholesterolemic activity. *Chem Nat Comp* 41:710–14.

Rho, S.J., Park, S., Ahn, C., Shin, J. and Lee, H.G. (2007). Dietetic and hypocholesterolemic action of black soy peptide in dietary obese rats. *J Sci Food Agric* 87:908–13.

Sacks, F.M., Lichtenstein, A.L., Van Horn, L., Harris, W., Kris-Etherton, P.K. and Winston, M. (2006). Soy protein, isoflavones, and cardiovascular health: an American Heart Association Science Advisory for Professionals from the Nutrition Committee. *Circulation* 113:1–11.

Santosa, S., Demonty, I., Lichtenstein, A.H., Ordovas, J.M. and Jones, P.J.H. (2007). Single nucleotide polymorphisms in ABCG5 and ABCG8 are associated with changes in cholesterol metabolism during weight loss. *J Lipid Res* 48:2607–13.

Setchell, K.D.R., Brown, N.M. and Lydeking-Olsen, E. (2002). The clinical importance of the metabolite equol – a clue to the effectiveness of soy and its isoflavones. *J Nutr* 132:3577–84.

Simon, J.S., Karnoub, M.C., Devlin, D.J., et al. (2005). Sequence variation in NPLC1L1 and association with improved LDL-cholesterol lowering in response to ezetimibe treatment. *Genomics* 86:648–56.

Sirtori, C.R., Eberini, I. and Arnoldi, A. (2007). Hypocholesterolemic effects of soya proteins: results of recent studies are predictable from the Anderson meta-analysis data. *Br J Nutr* 97:816–22.

Sugano, M., Yamada, Y., Yoshida, K., Hashimoto, Y., Matsuo, T. and Kimoto, M. (1988). The hypocholesterolemic action of the undigested fraction of soybean protein in rats. *Atherosclerosis* 72:115–22.

The UniProt Consortium (2007). The universal protein resource (UniProt). *Nucleic Acids Res* 35:D193–7.

Vigna, G.B., Pansini, F., Bonaccorsi, G., et al. (2000). Plasma lipoproteins in soy-treated postmenopausal women: a double-blind, placebo-controlled trail. *Nutr Metab Cardiovasc Dis* 10:315–22.

Wang, M., Yamamoto, S., Chung, H., et al. (1995). Antihypercholesterolemic effect of undigested fraction of soybean protein in young female volunteers. *J Nutr Sci Vitaminol* 41:187–95.

Yoshikawa, M., Fujita, H., Matoba, N., et al. (2000). Bioactive peptides derived from food proteins preventing lifestyle-related diseases. *BioFactors* 12:143–46.

Zhang, B., Chen, Y., Huang, L., et al. (2007). Greater habitual soyfood consumption is associated with decreased carotid intima-media thickness and better plasma lipids in Chinese middle-aged adults. *Atherosclerosis* 198:403–11.

Zhong, F., Zhang, X., Ma, J. and Shoemaker, C.F. (2007). Fractionation and identification of a novel hypocholesterolemic peptide derived from soy protein Alcalase hydrolysates. *Food Res Intl* 40:756–62.

Further reading

Almeida, S., Franken, N., Zandona, M.R., Osorio-Wender, M.C. and Hutz, M.H. (2005). Estrogen receptor 2 and progesterone receptor gene polymorphisms and lipid levels in women with different hormonal status. *Pharmacogenomics J* 5:30–4.

Anderson, J.W., Johnstone, B.M. and Cook-Newell, M.E. (1995). Meta-analysis of the effects of soy protein intake on serum lipids. *N Eng J Med* 333:276–82.

Choi, S., Adachi, M. and Utsumi, S. (2002). Identification of the bile acid-binding region in the soy glycinin A1aB1b subunit. *Biosci Biotechnol Biochem* 66:2395–401.

Nagaoka, S., Miwa, K., Eto, M., Kuzuya, Y., Hori, G. and Yamamoto, K. (1999). Soy protein peptic hydrolysate with bound phospholipids decreases micellar solubility and cholesterol absorption in rats and Caco-2 cells. *J Nutr* 129:1725–30.

CHAPTER

3

Business Applications of Nutrigenomics: An Industry Perspective

Rosalynn Gill

OUTLINE

Nutrition and Genomics
ISBN: 978-0-12-374125-7

SUMMARY

The completion of the Human Genome Project promised the development of many new technologies and discoveries, all based on the idea of releasing the potential within our genome. The application of genomic discoveries and technologies has led to the emergence of nutrigenomics, a scientific discipline focused on gene–diet interactions. By generating knowledge about the impact of nutrients on the genome and individual responses to nutrition interventions, nutrigenomics has opened new avenues for the industry. Building a successful nutrigenomics industry will, however, depend on scientific credibility, consumer interest and confidence and a clearly defined regulatory environment.

HUMAN GENOMICS AND THE POTENTIAL FOR BUSINESS APPLICATIONS

As the initial work of the Human Genome Project neared completion in the late 1990s and early 2000, the excitement was palpable in the biotechnology industry, the information technology sector and the public arena, too. What were the benefits of this massive undertaking going to be? How was understanding our 'genetic destiny' going to transform our daily lives, health care systems and the business world? Early conceptions of the potential applications of genomics fit logically into the pharmaceutical industry. Pharmacogenomics emerged as an ideal practical application of genetic information with analytical tools offering greater precision and lower costs to drug development and ultimately drug treatments.

The availability of genetic information and improved analytical tools has caused a revolution in nutrition research. Nutritional research is a field fraught with complexities and designing and carrying out effective basic nutritional studies present significant challenges to researchers. Food is a highly complex matrix, not a single chemical entity like a typical pharmaceutical agent. Human research subjects are difficult to control and, in most studies, the subjects are free living, consuming whatever and whenever they please and quantifying food consumption by intervention or by recall can be inaccurate. These challenges inherent in nutrition research are being overcome by technologies that allow for rapid, accurate and inexpensive measures of DNA, RNA, proteins and, now, metabolites. Nutrition scientists are able to incorporate analyses of genes and gene expression, which open a new understanding of how populations and individuals respond to dietary factors. By stratifying a study population based on genetic variations, including single nucleotide polymorphisms or insertions and deletions in the genetic code, researchers have been able to segment the study groups into categories that show differential responses to particular nutritional regimes, whether in terms of immediate biological effects via biomarker analysis, or through retrospective analyses in epidemiological studies.

There are many examples of the effect of genetic stratification on response to particular nutritional factors or potential disease risk. Two excellent examples may be found in Palli's studies examining the effects of GSMT1 and GSTT1 gene deletions on risk of lung cancer and on DNA adduct formation with various levels of consumption of fruits and vegetables (Palli, 2003, 2004). The studies are particularly illustrative because data are presented for all groups and then split into the different genotypes to demonstrate the impact of genetic variation of response. The applications of these advanced genomics technologies in nutritional research establish that, in nutrition, one size does not fit all. Further applications demonstrate the effects of bioactive compounds in food on the functioning genome by altering gene expression and changing metabolic profiles across populations. In this way, some bioactive compounds in food are shown to act in ways similar to pharmaceutical agents, thus blurring the boundaries between foods and drugs and introducing a dubious new phrase to the lexicon – 'phood'. Foods with scientifically demonstrated health benefits create opportunities to develop more efficacious functional foods.

LESSONS FROM FOOD INDUSTRY ACTIVITIES

To understand and predict developments in the nutrigenomics industry, it is worth drawing lessons about product development and consumer receptiveness to new products from other food industry experiences, including development of functional foods and expansion into health and wellness-based products. Functional foods with specific health benefits have been marketed for many years, including vitamin D fortified milk and folic acid enriched grains. Newer functional foods entering the marketplace have met with more challenges, particularly due to public resistance to genetically modified foods, regardless of whether or not genetic modification has been a part of a particular functional food's development.

The costs of research and development of new and more efficacious foods comes at a premium and represents a significant learning curve and potential cost barrier to an industry that has not typically carried out intervention trials in human subjects to determine response to particular ingredients and formulations. The food industry cannot command the margins typically found in the pharmaceutical industry, so R&D cost considerations are paramount for the food industry. Who will fund research to investigate the impact of food bioactives on both the individual and populations?

Who will bear the cost of systematic reviews of existing evidence to support the development of efficacious functional foods? Will consumers pay a premium for functional foods that have specific, proven health benefits? What are the costs of meeting regulatory requirements for marketing foods with health claims? These are critical questions for the food industry in considering applications of nutrigenomics research to new product development.

Consumers, especially the aging baby boomer generation, are increasingly interested in products and services that promote health and wellness, perhaps having been stimulated by newly created markets for these products and services. The health and wellness industry is growing rapidly and spans several categories, including the nutrition industry, the health food industry and the fitness industry. Major consumer companies, including Nestle, PepsiCo and Coca Cola are rebranding themselves as health and wellness companies and making strategic investments in products and services. For example, Nestle acquired Jenny Craig for $600 million, offering services and products focused on healthy weight loss programs. PepsiCo has leveraged its acquisition of Quaker Oats and its Gatorade brand for the marketing of healthy snacks and beverages. Finally, Coca Cola has acquired the maker of Vitamin Water for $4.2 billion to become a key player in the functional beverage industry.

THE GENOMICS IN NUTRIGENOMICS

Several consumer research surveys have identified specific consumer interests directly relevant to utilization of genetic information in nutrigenomics-based applications. In 2008, the Deloitte Center for Health Solutions reported that 61% of consumers want tools to provide recommendations to improve their health (Deloitte Center for Health Solutions, 2008). Cogent Research identified a population of over 100 million adults as 'very interested' in the use of genetic information to guide nutritional recommendations (Cogent Research, 2006). The International Food Information Council found that 79% of adult consumers had 'very' or 'somewhat favorable' attitudes towards the use of genetic information to provide diet or lifestyle recommendations (International Food Information Council, 2008). These results demonstrate that many consumers accept and expect that genetic information will one day be used to help determine food choices, particularly to promote health and wellness.

To address the question the individual consumer asks namely, 'What foods are right for me?', nutrigenomics companies have developed a genetic assessment model involving the use of genetic information of an individual to provide dietary and lifestyle advice. In this service delivery model, an individual submits a cell sample to a company for analysis and preparation of a testing report. The individual consumer's results may consist of a simple genetic report identifying particular genetic variations, a combined genetic report with nutritional and lifestyle advice and, in some instances, nutritional products like supplements may be offered.

Determination of the value chain is a major question in the evolution of this individualized approach to the business of nutrigenomics, also referred to as personalized nutrition or personalized genetics. That is, where does the money flow and how does this become a sustainable business model? There are several steps to this process, beginning with the genetic test. The test may be a single purchase enterprise, which then could become grouped into a 'diagnostic' type of business model. In this instance, rather than diagnosing a particular disease or condition, the diagnostic is to determine the right food, ingredient or nutritional supplement for the consumer based on genotype. The diagnostic market has typically been considered a low-margin, technically challenging business but, with consumer trends moving toward greater customization and personalization, there may be considerable growth in this industry. Still the question remains, how does a company build a sustainable business? Do companies sell as many tests to as many consumers as possible, or is there an opportunity for repeat sales as new genes, new discoveries and new food products and supplements become available?

Investors and analysts in the marketplace generally look for sustainable revenue sources and the diagnostic model holds little appeal unless it can lead to long-term repeat sales. For example, sales of a customized food product or supplement formulation based on a particular individual's genetic profile would be more appealing. This type of approach represents an attractive proposition if it means the introduction of premium products into a marketplace that has traditionally been considered a commodity market with low margins. Yet, challenges to the success of this type of model are many. From the perspective of business logistics, these challenges include:

- Genetic testing on a wide scale: currently the consumer pays out of pocket for assessments. Can the cost of assessment be reimbursable, who delivers the information to the consumer and how can this be carried out on a mass scale?
- Food ingredients and functional foods: can companies in the marketplace or governmental agencies bear the cost of research needed to validate the association of particular foods with particular genetic profiles?
- How is the food or ingredient delivered: are there a multitude of individualized formulae ('one size fits one'), or are there groupings of products that can be applicable to many consumers, but not all?

- The bottom line for the growth of the industry: is the consumer willing to pay a premium for genetically based assessments and for functional foods with the potential of conferring specific health benefits?

NUTRIGENOMIC TESTS IN COMMERCIAL APPLICATION

Current commercial nutrigenomic applications using genetic testing are generally focused on genes involved in metabolism of specific nutrients. For example, the most widely studied gene–diet or nutrigenomic association is that of the MTHFR gene and folic acid. The gene has a common single nucleotide polymorphism at position 677 in the coding region of the methylenetetrahydratefolate reductase (MTHFR) gene. The MTHFR enzyme is probably the most studied enzyme in the folate metabolism pathway. MTHFR directs folate from the diet either to DNA synthesis or to homocysteine remethylation. The C677T polymorphism of MTHFR results in the insertion of a valine residue in place of an alanine residue in the enzyme, which lowers both the stability and the activity of the enzyme as measured in cell extracts (Frosst et al., 1995).

There is substantial evidence that people with the C677T MTHFR polymorphism, which occurs at a relatively high frequency in the population (Ueland et al., 2001), have increased requirements for folate. These individuals tend to respond rapidly to folate supplementation with doses reported in the literature ranging from 0.2 mg to 5 mg (Tremblay et al., 2000). Examination of population frequency of the C677T polymorphism in various ethnic groups reveals a range of approximately 10% in African populations, 15% in African American populations, 35% in American Caucasian populations, up to 45% in Italian populations (Botto and Yang, 2000). The homozygous TT genotype occurs in approximately 10% of the Caucasian population (Botto and Yang, 2000).

The C677T polymorphism has been consistently linked with elevated homocysteine levels in blood plasma, also known as hyperhomocysteinemia (Jacques et al., 1996; Botto and Yang, 2000), especially when folate levels are low in the diet (Rozen, 2000). Several studies have shown that doses of folic acid in the range of 0.5 to 2.0 mg normalized the levels of homocysteine in 677TT individuals to a level equivalent with 677CC wild type individuals (reviewed in Ueland et al., 2001). Earlier studies have demonstrated that, in individuals with adequate folate status, the homocysteine levels are relatively low and do not vary between genotypes (Tokgozoglu et al., 1999).

For nutrigenomic testing applications in business, one model for assessment of folate requirements based on the C677T MTHFR polymorphism works as follows:

- An individual obtains a sample collection kit, either sold directly to the consumer or through a health care professional
- The individual submits a DNA sample – saliva, buccal swab or blood
- The DNA is analyzed for the presence or absence of the T allele at position 677
- If the T allele is found, the individual may be advised on the importance of maintaining adequate folic acid in his or her diet
- If data are available on the individual's current dietary habits, information on current consumption can be incorporated as well
- Finally, a folic acid dietary supplement or food enriched with folic acid may be recommended for the individual for a genetically personalized solution.

MARKET ACCEPTANCE OF NUTRIGENOMICS

The nutrigenomics industry will have to invest in research to build credibility and overcome skepticism regarding the possible efficacy of services and products. The industry will need to be confident that the investment will ultimately be returned as consumers and health professionals embrace nutrition based on scientific methodology. A critical factor facing the research community in nutrigenomics is the need to adopt rigorous standards and cohesive models for carrying out research and development, both in academia and in industry. The current 'gold standard' double blind, placebo-controlled model for pharmaceutical intervention needs to be revisited to accommodate the types of nutritional studies that can realistically be carried out by organizations or industries that have different economics and potential returns on investment other than the 'blockbuster' economics of the pharmaceutical industry. The methods of data capture, food intake analysis and, especially statistical analysis, need to be standardized and adopted in a way that will allow more accurate comparisons across studies and across populations (see Chapter 1).

Organizations such as the nutrigenomics Network of Excellence (NuGO) in Europe, have worked to bring together a network of academic and industrial researchers with the goal of adopting standardized procedures to help accelerate the pace of academic and industrial research (www.nugo.org). NuGO has fostered a spirit of collaboration by recognizing that many of the questions of nutrigenomics science are too large to be taken on by a single department or group and, by bringing together experts from academia and industry, to work toward solutions that can benefit the health of the public. Groups such as NuGO and the University of

California at Davis NCMHD Center of Excellence for Nutritional Genomics (www.nutrigenomics.ucdavis.edu) have been active in public communication of the concepts of nutrigenomics, but for consumers and health professionals to be aware of the potential and the limitations of nutrigenomics, educational efforts are needed on a much wider scale. Health care professionals, particularly dietitians and nutritionists, have an opportunity to play a key role in the development of this field, but these professionals will have to be educated on the underlying genetic information, which may prove a significant hurdle to acceptance and implementation by these groups.

Another key factor is the development of standards for utilization of information and associations from research in nutrigenomics. A view that is often repeated in the popular press is that 'the science isn't there yet' or 'it's too early', so establishing criteria for how much research is required before commercializing a genetic test would help the industry. Consider the case of the MTHFR gene, described above. Searching PubMed for the search string 'MTHFR and polymorphisms and folic acid' delivers 679 records for publications from 2000 to 2007. Is further research required? On the other hand, a PubMed search for literature on the APOA5 genetic variations, which have been associated with elevated triglycerides, delivered 70 records for 'APOA5 and polymorphisms'. When searching for a specific nutritional association, the term 'dietary fats' was added to the APOA5 search string and the number of records dropped to 3. A key factor for industries in this field is the determination of how much science is enough? As companies develop proprietary genetic associations, perhaps accompanied by specific food or supplement products, determining a threshold of research required will have a major impact on the cost and timing of development of new products or new assessments.

REGULATORY HURDLES FOR NUTRIGENOMICS

As with most new and developing technologies, the commercial applications of nutrigenomics have developed in advance of changes in the regulatory environment. The lack of clear regulations has led to considerable confusion and skepticism in the marketplace and the investment community. Ideally, the development of standards of practice within the developing industry could lead to greater consumer confidence. Establishing clear criteria for acceptance of scientific information and any health related information or claims that could be disseminated on the basis of this information, could remove many barriers to consumer and health professional acceptance. The relative newness of nutrigenomic science precludes the likelihood that government bodies can assemble such guidelines without input from subject matter experts, including participants from industry. Potential models discussed with regulators in the USA, for instance, include the development of a panel of subject matter experts, including a mix of academic scientists, health care professionals and industry experts to evaluate potential nutrigenomic applications.

A very real challenge for parties interested in bringing nutrigenomics to the marketplace is open communication and education of regulators who have been accustomed to more direct medical diagnostic applications for genetics, such as testing for cystic fibrosis or Huntington's chorea. Moving regulators and health care professionals from an expectation of a particular genetic result leading to a particular health outcome, as is the case with single gene disease diagnostics, to the more nuanced applications in nutrigenomics, where there is a significant dietary or environmental component that contributes to an expected outcome, has already proven to be a challenge for industry members. The challenges of moving from the deterministic 'gene equals disease equals diagnosis' model, coupled with the fact that the science of nutrigenomics is relatively young, with many of the studies based on the tracking of biomarkers, such as cholesterol levels, DNA adduct levels and homocysteine levels, requires a major paradigm shift in the traditional approach of medical and public health focused geneticists. The lack of disease-based outcomes from nutrigenomic research can lead to skepticism and dismissal in the regulatory community but, with a consumer population focused on taking proactive roles in preserving health and wellness, change in traditional methodology seems inevitable.

REGULATION OF GENETIC TESTING

There is considerable debate over which, if any, government entity should have oversight of the testing, particularly for testing for nutritional and health applications. In the USA, diverse federal regulations provide basic oversight on genetic testing proficiencies. This oversight currently is provided by the Centers for Medicare and Medicaid Services (CMS) under the Clinical Laboratory Improvement Amendments (CLIA) to ensure laboratory proficiency of genetic testing. Genetic testing laboratories applying for CLIA certification must meet standards for testing and quality control (Food and Drug Administration, 2007). The challenge for a CLIA inspector, however, is that there are many different techniques for measuring genetic variation, whether direct sequencing or various genotyping methods or platforms are used. Most laboratories develop genetic assay protocols in-house and the tests are thus true 'home brew' tests. A critical area for development of the industry is to establish coherent, cross-platform standards for validation and quality control of genetic testing, including chain of custody of individual samples. Many of the standards applied to traditional types of laboratory analyses, such as complete blood count tests, can be applied to genetic testing but, due to the variable nature of the techniques used for detection of genetic variations, the personnel charged with inspection must be able to apply standards that can be translated to the different circumstances of analysis.

Under CLIA, CMS is currently responsible only for the pre-analytic and analytic phases of genetic testing, with a particular interest in patient records and accessioning data, together with the actual laboratory results of testing. CMS is not currently charged with post-analytic evaluation, i.e. how the genetic information is translated or communicated to the end user, consumer or patient. The US Secretary's Advisory Committee on Genetics, Health and Society (SACGHS) has called for greater oversight of commercial genetic testing services, with the prospect of additional post-market monitoring by the Food and Drug Administration. SACGHS released a report in May 2008, 'US System of Oversight: A Response to the Charge of the Secretary of Health and Human Services', recommending a number of actions to strengthen CLIA oversight of genetic testing, and also recommended that the FDA address genetic testing 'in a manner that takes advantage of its current experience' (Secretary's Advisory Committee on Genetics, Health and Society, 2008). Of particular note, the SACGHS report urges that the level of FDA oversight of genetic testing be based on level of risk to the individual, thus recognizing that a 'one size fits all' regulatory approach to genetic testing would not be desirable.

Considering the broad range of genetic testing applications, from diagnostic tests for Huntington's chorea, to identification of breast and ovarian cancer risk associated with BRCA1 and BRCA2 mutations, and finally to the use of genetic testing to identify particular nutritional requirements for health, the recognition of the need for a tailored approach to the type of genetic testing appears to be a positive step in the development of a regulatory framework. This regulatory framework might support an emerging industry based on nutrigenomic applications and, at the same time, protect the interests of the consumer. The FDA released draft guidance documents in September 2006 on two specific issues relevant to genetic testing: one series addressed the use of commercially distributed analyte specific reagents (ASRs) used in developing laboratory tests, and the other series focused on a category of diagnostic genetic tests called *in vitro* diagnostic multivariate index assays (IVDMIA), which includes genetic tests and the algorithms used to obtain and deliver information that may lead to diagnosis or treatment of diseases (Food and Drug Administration, 2006). In response to the guidance documents, Sharon Terry from the Genetic Alliance, a group of stakeholders from industry, health care and consumers with interests in genetic testing, gave a public commentary on the SACGHS report (Terry, 2008). These comments pertained to:

- Improvement of oversight of genetic testing through the enforcement of regulatory authority under the CLIA program
- Registration of all genetic tests through a federal agency such as the FDA or the NIH, to include test performance and reference data, analytical and clinical validity and adverse event reporting
- Establishment of a laboratory-oriented consortium for sharing information related to quality control, validation and performance issues
- Creation of a public consortium of stakeholders to assess clinical utility and evidence related to genetic tests
- Agreement with the SACGHS report's concern over potential FDA oversight of information used in clinical applications
- Regulation of direct-to-consumer access to testing to ensure the public's safety
- Coordination by the US federal Department of Health and Human Services of relevant Health and Human Services agencies and interested stakeholders to develop a framework for genetic testing.

To complicate matters further in the USA, each state has its own set of regulatory guidelines for genetic testing, leading to a complex and often contradictory legal landscape. New York State, for example, has several requirements for licensure that have been broadly applied to all types of genetic testing. All genetic tests must be ordered by a health care provider, the results must be reported back to the health care provider and the patient must pay the laboratory directly for the genetic testing. The health care provider cannot bill the patient for the testing services. New York State has a detailed list of health care providers who are authorized to order the tests. Currently, dietitians and chiropractors are not authorized to order genetic tests but, for reasons that are unclear, podiatrists are authorized to order these tests. The New York State Department of Health has been active in the enforcement of their regulations and has issued Cease and Desist Letters to many companies involved in genetic testing, both for nutrigenomic applications and also for diagnostic or prognostic applications. The current restrictions on the method of ordering and paying for the tests represents a significant challenge for industries entering the field of genetic testing and the broad brush application of the regulations to all genetic tests at the time of this writing warrants further debate.

Should all genetic tests be considered medical or clinical in nature? The Department of Health from the State of California has recently followed the approach of New York State, issuing Cease and Desist Letters in June 2008 to companies involved in direct-to-consumer genetic testing. According to California statutes, laboratories that conduct clinical tests must be licensed. Again, the definition of what constitutes a clinical test is a matter of some debate in the industry. Nevertheless, at the present time, according to Code BP 1206 (a) (4) 'clinical laboratory test or examination' means the

> detection, identification, measurement, evaluation, correlation, monitoring, and reporting of any particular analyte, entity or substance within a biological specimen for the purpose of obtaining scientific data which may be used an as aid to ascertain the presence, progress, and source of a disease or physiological condition in a human being, or used as an aid in the prevention, prognosis, monitoring, or treatment of a physiological or pathological condition in a human being, or for the performance of non diagnostic tests for assessing the health of an individual (State of California, 2006).

California state law requires licensing for any laboratory that conducts tests of moderate or high complexity under CLIA on human tissue samples originating in the state of California (BP Code 1265). Genetic testing is considered high complexity testing under CLIA. Finally, California law (BP Code 1246.5) states that tests must be ordered by a physician. The Department of Health is currently working through the responses from the affected parties and has indicated that, in addition to ensuring the accuracy and quality of testing results, the State intends to examine post-analytic reporting, i.e. the scientific substantiation of the information that is communicated following the genetic test. In both California and New York, the timeline for evaluation and the resources available to examine and grant licensure is a significant issue. For companies that wish to become licensed in the states, licensure can take months to complete, causing a significant impediment to business development in the two largest markets in the USA.

In 2003, in the UK, the Human Genetics Commission (HGC) carried out a lengthy consultation with the public and with specific stakeholders to provide recommendations to the government on the development of a clear regulatory strategy (Human Genetics Commission, 2003). Of particular relevance to the business of nutrigenomics was the conclusion of the HGC that there is a spectrum of genetic tests, so regulations should be developed that can accommodate tests with diverse purposes. A different set of criteria for a test diagnosing Huntington's chorea, for instance, would be needed in comparison with a nutrigenomic test identifying requirements for a particular nutrient. The HGC released another report in December 2007 (Human Genetics Commission, 2007). One of the recommendations included in this report was the development of a code of practice relating to genetic testing services supplied directly to the public. The HGC continues to consult with stakeholders and is exploring the adoption of voluntary regulation of direct-to-consumer (DTC) tests through a series of discussions with the goal of reaching a consensus on standards of good practice that could be adopted by companies involved in DTC tests.

In 2006, companies involved in DTC nutrigenomic testing to the public came under scrutiny by the Senate Special Committee on Aging, with the release of a report by the US Government Accountability Office (US Government Accountability Office, 2007). Representatives from industry, including the author of this chapter, were called to testify to the committee. The testimony of the participants and a rebuttal of the GAO report, written by the author of this chapter, now form part of the congressional record. Experience with the GAO investigation revealed how important open dialogue and clear understanding will be to the development of a regulatory framework for nutrigenomic applications because early and effective communication can dispel misperceptions about nutrigenomics. One of the key conclusions of the GAO report was that consumers were being misled and that genetic information was not used in preparation of consumer reports. The investigator described the use of 14 different lifestyle questionnaire answers resulting in 14 different report outcomes. In fact, the questionnaires are used to provide personalized information on dietary and lifestyle habits and this information is presented alongside genetically based information. The 14 different questionnaires should produce personalized feedback related to the questionnaires juxtaposed with genetically based goals. The GAO report does not disclose that the 14 different reports gave identical genetic information for the samples that were submitted, a result that indicates the laboratory tests conducted by different companies were providing consistent results.

Representatives from government regulatory agencies and the academic community were also asked to testify at the Senate hearing. A positive result of this debate has been greater dialogue between industry participants and regulatory representatives. A key factor in discussions with the regulators, similar to that with the HGC in the UK previously, has been the determination of perceived 'risk' to the individual. In general, there is consensus currently that the use of genetic testing to provide dietary and lifestyle advice represents a relatively low risk to the individual, but concerns remain over the development of appropriate standards for genetic analysis and communication of information.

PRIVACY AND DISCRIMINATION PROTECTION FOR NUTRIGENOMIC APPLICATIONS

Protection of privacy of an individual's personal information, genetic or otherwise, will be essential to the success of companies using this information for commercial applications of nutrigenomics. In the USA, the Genetic Information Non-Discrimination Act was enacted in 2008 (United States Congress, 2008) to prohibit insurance companies and employers from discriminating against individuals on the basis of their genetic information.

Consumers express strong concerns regarding privacy of their personal information, especially when dealing with companies that offer DTC nutrigenomic services. In the USA, the Health Information Portability and Accountability Act (United States Congress, 1996), provides guidelines on the use and handling of personal data. The development of industry standards of practice for the storage, protection and transmission of personal genetic information would be of great benefit to the industry.

RECOMMENDATIONS FOR IMPROVING THE NUTRIGENOMICS BUSINESS ENVIRONMENT

Those wishing to improve the business environment for nutrigenomics ought to consider two initiatives: establishing an industry group, and incorporating an ethical framework into the business of nutrigenomics.

Establishing an Industry Group

Establishing an industry group to assist in the development of a regulatory framework will be critical for the success of business opportunities in nutrigenomics. There is little expertise or awareness of nutrigenomic science in government agencies around the world and, because the science and commercial applications are so new, there are considerable challenges in informing and educating regulatory authorities. Whether discussing the introduction of functional foods, in which pharmaceutical models may not be appropriate either mechanistically or economically, or discussing the nuanced outcomes expected from interventions or testing based on gene–diet or gene–environmental associations, there is considerable need for education regarding the current state of the science and the potential commercial applications of nutrigenomics.

Incorporating an ethical framework into the business of nutrigenomics

Business models incorporating nutrigenomics, whether through actual genetic testing of individuals, or through the use of functional ingredients with potential health benefits, must address many potential ethical issues. There are several key factors. To begin, there is a question about who actually owns the genetic information of a particular individual. Surprisingly, this fundamental issue has been largely unexplored in the public arena. Genetic data contained in databases and biobanks can be used to discover new associations with disease, with lifestyle or dietary habits. It is essential clearly to define who can have access to the information and how the individual can ensure access to or perhaps restrict his or her own genetic information. Second, and related to the first, do individuals have rights to their genetic information? As the stakeholders in the genetic testing industry struggle to find a clear regulatory path, the right of the individual to have access to genetic information needs to be explored. Some companies have adopted policy statements to the effect that each individual has the right to his or her own genetic information (see www.sciona.com or www.navigenics.com for examples of these policy statements). These statements do not, however, reflect a universal or generally accepted standard at this time. Third, how is the information communicated to individuals? Whether providing genetic information or describing potential health benefits of functional ingredients, what standards should be developed to ensure responsible and accurate communication of information? In the USA, the Federal Trade Commission regulates advertising practices but the development of industry standards for nutrigenomics could provide a framework that benefits all parties, builds consumer trust and helps regulators gain greater understanding of the benefits and limitations of nutrigenomic applications. Fourth, how should information be obtained? DTC tests provide an alternative to having health care professionals as conduits of genetic information. A survey of stakeholders in the UK revealed a preference for restricted access to the information, a view the authors describe as 'institutionalized paternalism' which prevents access to valuable health information and stifles individuals from taking a more proactive role in their health (Carter et al., 2006).

THE FUTURE FOR THE BUSINESS OF NUTRIGENOMICS

What does the future hold? With advances in science, better understanding from various regulatory authorities all over the world and with a concerted effort from stakeholders from industry, interested consumer groups and governmental representatives, the future seems filled with possibilities for the development of successful business models in nutrigenomics. Beyond the corporate bottom line, nutrigenomics can make significant improvements to the health and well-being of the consumer.

What might the future look like? A preliminary study of nutrigenomics in clinical practice was published recently in the *Nutrition Journal* (Arkadanios et al., 2007). In this study, patients on a clinical weight management program were offered a nutrigenomic analysis as part of their treatment regime. Results of the genetic analysis led to modifications to a platform dietary program, such as addition of folic acid when the MTHFR C677T variation was identified. When compared against a matched, untested population over an extended period, the tested population achieved better outcomes in terms of maintaining body mass index reductions over a one-year period as well as greater reductions in blood glucose levels. Analysis of additional biomarkers is currently underway, however, initial trends indicate that the tested population had greater results in terms of homocysteine reduction and LDL cholesterol reductions. While the study population was relatively small – 50 tested subjects versus 43 untested subjects – this paper is the first documentation of nutrigenomics in practice and demonstrates positive results from the use of genetic information in developing dietary guidelines for individuals.

Nutrigenomics is a business with significant challenges and hurdles to overcome, but there is widespread recognition in the public, in the academic world and in the industrial world that the use of genetics to guide dietary and lifestyle choices is going to happen, indeed, the business of nutrigenomics already is happening as reported in the study above. Questions remain about the execution of different aspects of business models, how businesses can expand to accommodate the population at large, how quickly the science will mature and how regulations can be developed to protect the consumer but not impede the industry. The answer: cooperation between stakeholders is absolutely essential. All the relevant stakeholders must engage in open dialogue about the issues so that the ultimate promise of nutrigenomics – the promotion of health for the public through greater understanding of the genetic response to nutrients – can be achieved.

References

Arkadanios, I., Valdes, A.M., Efstathios, M., Anna, F., Gill, R.D. and Grimaldi, K.A. (2007). Improved weight management using genetic information to personalize a calorie controlled diet. *Nutr J* 6:6–29.

Botto, L.D. and Yang, Q. (2000). 5,10-Methylenetetrahydrofolate reductase gene variants and congenital anomalies: a HuGE review. *Am J Epidemiol* 151:862–77.

Carter, S., Taylor, D. and Bates, I. (2006). Institutionalized paternalism? Stakeholders' views on public access to genetic testing. *J Health Serv Res Pol* 11:155–61.

Cogent Research (2006). Cogent syndicated genomics attitudes and trends: 10–19.

Deloitte Center for Health Solutions (2008). 2008 Survey of health care consumers executive summary: 5–20.

Food and Drug Administration (2006). Draft guidance for industry, clinical laboratories, and FDA staff – in vitro diagnostic multivariate index assays. CDRH, Bethesda.

Food and Drug Administration (2007). Clinical Laboratory Improvement Amendments (CLIA), 42 USC 263a (http://www.fda.gov/cdrh/clia/).

Frosst, P., Blom, H.J., Milos, R., et al. (1995). A candidate genetic risk factor for vascular disease: a common mutation in methylenetetrahydrofolate reductase. *Nat Genet* 10:111–13.

Human Genetics Commission (2003). Genes direct – Ensuring the effective oversight of genetic tests supplied directly to the public. Department of Health, London.

Human Genetics Commission (2007). More genes direct: a report on developments in the availability, marketing and regulation of genetic tests supplied direct to the public. Department of Health, London.

International Food Information Council (2008). 2007 Consumer attitudes towards functional foods/foods for health executive summary: 1–16.

Jacques, P.F., Bostom, A.G., Williams, R.R., et al. (1996). Relation between folate status, a common mutation in methylenetetrahydrofolate reductase, and plasma homocysteine concentrations. *Circulation* 93:7–9.

Rozen, R. (2000). Genetic modulation of homocysteinemia. *Semin Thromb Hemost* 26:255–61.

Secretary's Advisory Committee on Genetics, Health and Society (2008). US system of oversight of genetic testing: a response to the charge of the secretary of health and human services. Report of the Secretary's Advisory Committee on Genetics, Health, and Society. NIH, Bethesda.

State of California (2006). Business and Professions Code. http://www.leginfo.ca.gov/calaw.html. Accessed July 28, 2008.

Terry, S.F. (2008). Comments on behalf of the Board of the Genetic Alliance Secretary's Advisory Committee on Genetics, Health and Society. http://geneticalliance.org/policy.genetic.testing.oversight.

Tokgozoglu, S.L., Alikasifoglu, M., Unsal, A.E., et al. (1999). Methylene tetrahydrofolate reductase genotype and the risk and extent of coronary artery disease in a population with low plasma folate. *Heart* 81:518–22.

Tremblay, R., Bonnardeaux, A., Geadah, D., et al. (2000). Hyperhomocysteinemia in hemodialysis patients: effects of 12-month supplementation with hydrosoluble vitamins. *Kidney Int* 58:851–8.

Ueland, P.M., Hustad, S., Schneede, J., Refsum, H. and Vollset, S.E. (2001). Biological and clinical implications of the MTHFR C677T polymorphism. *Trends Pharmacol Sci* 22:195–201.

United States Congress (1996). Health Insurance Portability and Accountability Act of 1996. Public Law No. 104-191,110 Stat 2033. http://www.cms.hhs.gov/HIPAAGenInfo/Downloads/HIPAALaw.pdf. Accessed July 28, 2007.

United States Congress (2008). Genetic Information Non-Discrimination Act (GINA) of 2008. H.R. 493. http://www.govtrack.us/congress/billtext.xpd?bill=h110-493. Accessed July 28, 2008.

Further reading

Palli, D., Masala, G., Vineis, P., et al. (2003). Biomarkers of dietary intake of micronutrients modulate DNA adduct levels in healthy adults. *Carcinogenesis* 24:739–46.

Palli, D., Masala, G., Peluso, M., et al. (2004). The effects of diet on DNA bulky adduct levels are strongly modified by GSTM1 genotype: a study on 634 subjects. *Carcinogenesis* 25:577–84.

US Government Accountability Office (2006). Nutrigenetic tests: tests purchased from four websites mislead consumers. GAO, Washington, DC.

4

Regulation of Genetic Tests: An International Comparison

Stuart Hogarth

OUTLINE

Nutrition and Genomics
ISBN: 978-0-12-374125-7

SUMMARY

This chapter discusses regulatory frameworks for genetic tests, in particular the systems for regulating medical devices and clinical laboratories. It describes concerns about genetic testing that have led to calls for enhanced regulation and considers recent developments in regulation. It identifies features of regulatory systems in several jurisdictions, including the UK, the USA, the European Union, Canada and Australia. The chapter concludes by discussing several different policy options that are emerging both from developments in regulatory systems and policy debates surrounding genetic testing. These options include: pre-market review and post-market controls; genetic test and laboratory registries; enhanced roles for third parties; and consumer education.

INTRODUCTION

How should we regulate genetic testing? This question has troubled clinicians, patient groups, policy makers and regulators for over 10 years. In this chapter, regulatory frameworks for genetic tests are reviewed, in particular the systems for regulating medical devices and clinical laboratories. Recent developments in regulation are considered in light of long-running policy discussions about how to enhance regulation of genetic tests. The chapter concludes by setting out a number of different policy options that are emerging both from developments in regulatory systems and policy debates. This chapter does not address the specific challenges associated with the direct-to-consumer provision of genetic testing, which are considered in Chapter 5. Before providing an overview of regulations applicable to genetic testing, the context is set by outlining concerns about genetic testing that have led to calls for enhanced regulation. A broad definition of regulation is adopted that includes traditional 'command and control' instruments of statutory regulation and more diffuse control mechanisms that are sometimes encompassed in the concept of 'governance', but which are also covered by the concept of 'de-centered regulation'.

CONCERNS ABOUT GENETIC TESTS

Some nutrigenetic tests have been criticized by government bodies, scientists, doctors and consumer organizations in the UK (GeneWatch UK, 2002; Human Genetics Commission, 2002) and the USA (US Government Accountability Office, 2006; US Congress, 2006). These criticisms need to be placed in the context of broader concerns about genetic testing. Over the past 15 years, as genetic testing has begun to play a greater role in disease prevention, management and treatment, there have been growing concerns about the potential for harm arising from the lack of a robust governance framework for genetic testing.

Few, if any, of the harms associated with genetic tests are unique to genetic tests. Nevertheless, genetic tests have raised particular concern for a number of reasons, not least of which is last century's legacy of eugenics and fear of a new eugenics based on genetics. The clinical impact of genetic tests also gives cause for concern, particularly when they are used for reproductive decision-making, or where a test may have significant psychological impact because it predicts the likelihood of an incurable or serious disease. In the case of tests for common diseases, there is a concern that overuse of susceptibility tests of limited predictive value will expand the class of the worried well. Predictive tests (genetic and non-genetic) raise other issues. For instance, they provide an uncertain guide to the likelihood of disease, its severity and the timing of onset (Evans et al., 2001). Doctors cannot learn from experience with predictive tests because they cannot compare test results with the patient's symptoms. By definition, such tests are 'stand-alone,' meaning there is no other testing method that can be used to confirm the test result (Foucar, 2001). In the

TABLE 4.1 ACCE framework

Analytic validity–accuracy of test identifying the gene
Clinical validity–relationship between the gene and clinical status
Clinical utility–likelihood that test will lead to an improved outcome
Ethical, legal and social implications

case of nutrigenetic testing, particular genes may be associated with fast or slow metabolism of particular nutrients and thus increased risk of specific diseases. However, that neither means that an individual who has those genes will develop the disease nor that an individual who does not have those genes can be reassured they will remain disease-free.

One key concern has been that some genetic tests are entering clinical practice prematurely. As one senior diagnostics industry figure put it:

> [There has been] a noticeable lack of consensus within the genetics community about exactly when a test for a new marker was sufficiently validated for it to enter into clinical service. Some labs rushed to provide testing after the first publication, while others waited until the result had been replicated in multiple studies or multiple ethnic groups (Winn-Deen, 2003).

Some fear that the premature commercialization of poorly validated tests will increase as tests for common diseases become more widespread (Khoury et al., 2004). Critics, including clinical geneticists, argue that poor quality tests are being rushed to market, especially in the area of predictive and susceptibility testing and nutrigenetic testing (Haga et al., 2003). As a consequence, there have been repeated calls for more systematic pre-market review of tests. These concerns focus on whether tests are being used before their clinical validity and clinical utility have been clearly established, however, the analytic validity of tests has also been a concern (see ACCE framework, Table 4.1). A recent US study looking at quality assurance of laboratory testing procedures revealed wide variations in laboratory performance measured by the number of deficiencies in formal proficiency testing and the number of incorrect test results reported by the laboratory (Hudson et al., 2006). A number of reports have expressed concern about the lack of, or variations in, controls over laboratory quality assurance (Ibaretta et al., 2003; OECD, 2007a).

Finally, there are concerns about the quality of information patients and doctors receive pre- and post-test. These concerns focus on information provided in a variety of ways: through pre- and post-test counseling; printed or online literature produced by those offering tests; and the use of direct-to-consumer advertising and other consumer marketing channels to promote genetic tests. A 2002 study showed significant variation in the way molecular genetic testing laboratories report results for cystic fibrosis and factor V Leiden testing (Andersson et al., 2002). In 2003, a follow-up study reported the results of a physician survey that indicated simple, yet comprehensive and useful genetic test result reports were sought by physicians with the implication that such reports are often not available (Krousel-Wood et al., 2003).

The primary regulatory challenges associated with genetic testing include the pre-market review of tests to ensure they are fit for purpose (i.e. can diagnose or predict disease with the accuracy that the test developer claims), quality assurance of laboratory procedures to ensure accurate testing and the policing of promotional claims to ensure consumers are not misled. Other challenges include protecting the privacy of genetic data and ensuring that informed consent takes place.

REGULATORY FRAMEWORKS

Having identified some key regulatory challenges, existing regulatory mechanisms that might address these challenges, specifically regulatory frameworks in Europe and North America for medical devices and clinical laboratories are now described. The assumption is that these regulations cover nutrigenetic tests. There has been some doubt about this issue, at least in the UK, with the regulatory body for *in vitro* diagnostic (IVD) tests, the Medicines and Healthcare Products Agency (MHRA), claiming that because so-called 'lifestyle' tests are not clinical tests, they would not be covered by the relevant legislation (Human Genetics Commission, 2007). The term 'lifestyle test' is not precisely defined, but is sometimes used to describe nutrigenetic tests where the intent is to provide lifestyle advice such as dietary guidance. The MHRA draws a distinction between these and what it deems 'tests for a medical purpose'.

What constitutes an IVD medical device is an important issue. In Europe, a medical device is an item 'intended by the manufacturer to be used for human beings, for the purpose of …. diagnosis, prevention, monitoring, treatment or alleviation of disease' (European Commission, 1993) and the definition of an '*in vitro* diagnostic medical device' is any medical device which is '. . . intended by the manufacturer to be used in vitro for the examination of specimens . . . derived from the human body, solely or principally for the purpose of providing information: concerning a physiological or pathological state . . .' (European Commission, 1998).

The Human Genetics Commission (HGC) has questioned the MHRA's position and suggests that 'lifestyle' tests may be considered IVD devices if their purpose is to help in the prevention of disease (Human Genetics Commission, 2007). Moreover, a number of regulatory authorities have indicated they agree with the HGC. At a congressional hearing largely devoted to the regulation of nutrigenetic tests, Steve Gutman, Director of the US Food and Drug Administration's (FDA) Office of *In Vitro* Diagnostics, answered in the affirmative when asked whether such tests were covered by FDA's regulations for medical devices (US Congress, 2006). Letters were subsequently sent to a number of nutrigenetics companies inviting them to meet with the FDA. In Australia, the Therapeutic Goods Administration (2007) has issued guidance on nutrigenetic tests.

REGULATION OF MEDICAL DEVICES

Genetic tests fall under broader statutory regimes for the regulation of medical devices because they are *in vitro* diagnostics. Although generally less burdensome than the regimes for pharmaceutical products, they share a number of key features: they are concerned with ensuring the safety and efficacy of health care products and their most powerful tool is their authority both to grant new products permission to enter the market and to remove existing products from the market should serious problems arise.

In the USA and Europe, failure to comply with regulations is a criminal offence (US Congress, 1994). (For an overview of how criminal sanctions vary across EU member states, see Pilot, 1999.) Device regulators use a risk-based approach to determine which categories of products require stringent scrutiny and which can be dealt with more leniently. The fundamental elements of the regulatory system are: registration with the regulatory authority; quality assurance; pre-market review; truth-in-labeling; and post-marketing surveillance.

Registration

Registration requires that manufacturers provide the regulator with basic details of their organization and also of the products they are placing on the market. It is a minor requirement in itself, but provides the regulator with essential information should it be called upon to investigate a problem with a device.

Quality assurance

Device manufacturers are obliged to follow quality assurance (QA) systems. Manufacturing facilities are subject to periodic inspection by government or accredited third party agencies. Manufacturers can meet QA requirements through certification to recognized standards, such as those developed by the International Organization for Standardization (ISO), in particular ISO 13485, which sets out quality systems standards for use in the regulation of medical devices. Quality system requirements can cover everything from initial design through to post-market vigilance. Quality assurance in the manufacturing process is often referred to as good manufacturing practice (GMP).

Pre-market review

Before placing their tests on the market, manufacturers must ensure they meet the requirements set out by the regulator. Pre-market review of the product and the quality assurance system to ensure they are acceptable is a process known as conformity assessment and involves the systematic examination of evidence generated and procedures undertaken by the manufacturer. For low risk devices, the manufacturer will carry out this review themselves. For moderate and high risk devices, the review is carried out by the regulator or a third-party body empowered to act for the regulator.

Truth-in-labeling

Pre-market review is used to ensure truth-in-labeling, i.e. that the manufacturer's intended use for the product is supported by clinical data on the test's performance as set out in the technical file and summarized in the product label and in promotional material. The product label will also include instructions for the user including any necessary warnings on the limitations of the test's performance. Ensuring truth-in-labeling and truthful promotion—an honest account of the strengths and weakness of a test's performance—can be thought of as the fundamental function of pre-market review in the medical devices sector, although for high risk tests the process may be more onerous, with regulators setting out in some detail the types of clinical studies required to gain pre-market approval.

Post-marketing controls

Once a device is on the market, it is subject to post-marketing controls. Regulators can place restrictions on the sale, distribution or use of devices and remove unsafe products from the market. In the past, device regulation, like drug regulation, has tended to focus on pre-market review, but post-marketing surveillance has taken on increasing importance in recent years. For instance, in Europe, manufacturers are required to have a systematic procedure to review experience gained from their devices in the post-production phase. European guidance indicates the importance of post-marketing studies in certain circumstances—such as the severity of the disease or the novelty of the technology—and indicates the range of approaches to data collection which can include 'extended follow-up of patients enrolled in the pre-market trials, and/or a prospective study of a representative subset of patients after the device is placed on the market. It can also take the form of open registries' (European Commission, 2004).

LIMITATIONS OF CURRENT REGIMES FOR REGULATING GENETIC TESTS AS MEDICAL DEVICES

Laboratory-developed tests and medical device regulations

Genetic testing is characterized by a high degree of dependence on laboratory-developed tests (LDTs). In the European Union, Sweden and Australia, LDTs are included in the device regulations (although there are exemptions in the EU system for laboratories based within health institutions). In Canada, device regulators have sought legal opinion on whether they can regulate LDTs and have received a succession of conflicting opinions. Recently, they have stated that the development and use of an LDT within a health care facility does not constitute a 'sale' as defined in the Canadian *Food and Drugs Act* since legal ownership of the LDT does not change. Should the health care facility share their proprietary LDT with other health care facilities so that they may test samples, however, it would constitute a 'sale' and the LDT would be subject to the applicable provisions of the Act and Medical Devices Regulations (Hogarth, 2007).

The FDA has vacillated on the issue of whether they have the authority to regulate in-house tests but, in the last 2 years, the agency has gradually begun to intervene on a case-by-case basis. This piecemeal approach has now culminated in a draft FDA guidance in which the agency asserts it has authority to regulate in-house tests as medical devices and indicates it is now intending to exercise that authority over a class of complex tests that require interpretative algorithms to generate results (referred to as IVDMIAs—*in vitro* diagnostic multivariate assays) (FDA, 2006a).

While Europe treats commercial LDTs as devices subject to the IVD Directive, it is not clear that this applies to LDTs performed by laboratories outside Europe. For instance, the US companies InterGenetics and Myriad have both made their tests available through third parties in the UK; others, such as Genomic Health, are following suit. These UK third parties collect the samples and return the results to consumers, but the test is performed by the company in the USA in their own reference laboratory. The regulatory status of such tests is currently unclear. Were such US companies to be exempted from the Directive, it may place European LDT companies at a commercial disadvantage. For instance, the Dutch company Agendia, whose MammaPrint test is the main competitor of Genomic Health's Oncotype Dx, not only needed a CE mark for their test in Europe, but had to gain FDA approval to market their test in the USA. (A CE mark—*'Conformité Européene'*—denotes compliance with relevant European regulatory directives.)

The provision of information is another area where there is a clear difference between device regulation and the regulation of LDTs. Even where they have been deemed medical devices, there is currently no regulatory equivalent of a label for LDTs. Although laboratory regulation covers interpretation of test results and communication of results to doctors/patients, this is at the post-test stage. These regulatory regimes do not address the pre-test stage, where doctors and their patients are deciding whether to use a test. Furthermore, there are no statutory regulations on what types of data should be in a test results report (although this is addressed in professional guidelines such as those produced by the College of American Pathologists, 2007).

This lack of control over the provision of information about LDTs becomes particularly problematic when such tests are sold direct-to-consumer (DTC). IVD regulations do acknowledge that DTC testing kits pose specific risks that merit greater regulatory scrutiny. For instance, in Europe, self-testing kits (e.g. glucose meters) are reviewed by a notified body to ensure they can be understood and used easily by the lay public. The Directive states that:

> the results need to be expressed and presented in a way that is readily understood by a lay person; information needs to be provided with advice to the user on action to be taken (in case of positive, negative or indeterminate result) and on the possibility of a false positive or false negative result (European Commission, 1998).

This higher level of regulatory scrutiny does not apply to a DTC testing service, despite the fact that the consumer is no better prepared to understand the test results than if they had purchased the test as a kit. This is another example of the uneven playing field between kits and LDTs.

Risk classification

As noted earlier, regulatory regimes for medical devices are based on risk classification. Tests considered higher risk, because of their clinical or public health significance, or sometimes their novelty, are subject to greater scrutiny. Regulatory gaps can appear when tests are deemed low risk and therefore exempt from independent pre-market review.

In the USA, Canada and Australia, genetic tests that fall within the medical device regulations are all treated as moderate to high risk and so are generally subject to pre-market review (although in Australia some genetic tests are Class II and exempt from pre-market review). Europe differs from these other countries because (with a single exception), genetic tests are

TABLE 4.2 Risk classification

Country/region	Risk categories	Genetic test risk classification	Pre-market review
USA*	I–III	II or III	Yes
Canada**	I–IV	III	Yes
Australia**	I–IV	II or III	Yes (class III)
Europe*	I–III	I (except PKU)	No

*3 class system: I = low risk; II = moderate risk; III = high risk.
**4 class system: I = low risk; II = low to moderate risk; III = moderate to high risk; IV = high risk.

treated as low risk under the IVD Directive and are exempt from independent pre-market evaluation (Table 4.2). Instead, the manufacturer decides if it has fulfilled its obligations under the Directive and, having done so, adopts the 'CE' mark.

This reflects a wider problem with the European framework. In effect, it lacks a coherent mechanism for classifying the risk profile of new tests. The vast majority are considered low risk, with only a small number classed as moderate or high risk (certain blood screening tests). There is little consistency in this scheme regarding moderate risk classifications. Chlamydia tests, for example, are moderate risk, but other sexually transmitted disease tests are low risk. Prostate-specific antigen (PSA) testing, a screen for prostate cancer, is moderate risk, but other cancer tests are not, and tests for the heritable disorder phenylketonuria (PKU) are moderate risk, but all other heritable conditions are low risk. The *prima facie* assumption is that all new tests are low risk and the mechanism for adding tests to the high or moderate risk category has been used only once.

Post-marketing surveillance

Despite an avowed emphasis on post-marketing controls in device regulation, it would seem that post-marketing surveillance has not changed much in practice for either regulators or manufacturers. For the regulators, institutional capacity might be the main problem (Altenstetter, 2005). Effective post-marketing surveillance needs effective ways to collect and share data on the performance of devices once they are on the market. In the USA, the FDA has recently begun to address this problem in a new initiative designed to enhance their post-marketing activities (FDA, 2006b). In the context of limited capacity for effective post-marketing activities, regulators may rely on a continued emphasis on pre-market review, however, in the European context, where practically no tests are subject to independent pre-market review, the failure to develop an adequate system of post-marketing surveillance is more critical.

REGULATION OF CLINICAL LABORATORIES

There is an international trend toward the regulation of LDTs as medical devices, but this is not the only regulatory regime governing the work of clinical laboratories providing genetic tests. Laboratories are also subject to a range of licensing and accreditation controls. In general, regulation of clinical laboratories is quite distinct from device regulation and is focused on quality assurance of laboratory procedures and the analytical accuracy of laboratory testing which is monitored through periodic laboratory inspections.

Registration

As with medical device regulations, the most basic form of control exerted over clinical laboratories is the requirement to register as an organization with the licensing or accrediting authority.

Quality assurance

Regular inspections take place to ensure that the correct procedures are in place and are being correctly followed. As with medical device regulation, many aspects of laboratory quality assurance can be performed using internationally agreed standards such as ISO 17025, a general standard for testing or calibration services, or ISO 15189, a standard specific to medical laboratories. Proficiency testing is the most rigorous means to assess the accuracy of laboratory testing as it involves comparing a laboratory's test performance and results to an established external standard. Quality assurance may also cover issues such as the provision of information to doctors and patients, as well as the need to maintain records.

Pre-market review

While laboratory regulations can require that laboratory directors verify and establish the test's analytical performance characteristics, the only regulatory regime for clinical laboratories which includes pre-market review of LDTs is in New York State. Under the US federal *Clinical Laboratory Improvement Amendments* (CLIA), individual states can have their own system for laboratory licensing, with requirements stricter than the CLIA regulations. Laboratories licensed by New York State are required to submit clinical validity data on new tests for pre-market approval (New York State, 2007). There is very little difference between the data submission for approval of a LDT by New York State and a Class II submission for FDA. This system has a major impact on genetic testing in the USA because all the major reference laboratories, and many of the medium-sized ones, are New York State-licensed (Willey, 2007).

Result reporting

Laboratory regulations may set out the types of information that should be provided to the person who has ordered the test, such as reference ranges, interpretation and issues that may affect the accuracy of the test.

Personnel

Laboratory regulations may stipulate the level of qualification required by staff conducting tests and may stipulate in some detail the responsibilities of senior staff such as the laboratory director.

LIMITATIONS OF CURRENT REGIMES FOR REGULATING CLINICAL LABORATORIES PERFORMING GENETIC TESTS

Lack of proficiency testing

Proficiency testing is not mandatory in the USA. While CLIA requires laboratories to have quality assurance programs in place, most genetic testing laboratories are not required to perform the type of assessment called proficiency testing unless they are testing a small subset of established analytes regulated under CLIA, none of which are genetic tests *per se*.

Regional/national variation in regulation

There are significant variations in the level of control exerted over clinical laboratories. In the USA, clinical laboratories are governed by CLIA but, in Europe, compulsory licensing of all clinical labs is rare. Many European countries have voluntary accreditation schemes instead. Even within a single country, considerable regulatory variation can exist. In the USA, the New York State regulations offer a significantly higher level of protection than the basic CLIA system. A similar variation exists within Canada where laboratory regulation is a matter for individual provinces. Some, but not all, have statutory regulations in place, some have a licensing system and others operate an accreditation system (Petit et al., 2008). A number of national bodies have a role including the Standards Council of Canada, the Canadian Council on Healthcare Services Accreditation (in conjunction with the Canadian Standards Association) and the Canadian College of Medical Geneticists, which issues its own guidelines for quality assurance.

The lack of common European requirements for laboratory quality assurance may arise because laboratory testing is generally carried out within national healthcare services, a sphere of activity that is seen as the responsibility of member states. A 1997 survey suggested that, in EU countries where clinical genetics is well established, there was often a legal framework governing the service (Harris and Reid, 1997). A more recent survey of European countries found that seven had legislation: Austria, Belgium, France, Norway, Sweden, Switzerland and the Netherlands (Godard et al., 2003). This legislation often covers the licensing of laboratories but sometimes relates to the clinics within which genetics is practiced. Even where such legislation exists, however, it may not mandate formal quality assurance and proficiency testing. A recent Europe-wide survey revealed that very few laboratories have formal accreditation – up to 50% of laboratories surveyed do not undergo any official inspection (Ibaretta et al., 2003).

Focus on analytic performance

The primary focus of laboratory quality assurance is the analytic accuracy of testing. Although there may be provisions relating to issues such as the provision of information prior to testing and the delivery of test results, these are not the main focus of laboratory quality assurance and, as a consequence, there are no clear standards for laboratories to achieve.

Lack of transparency

There is a lack of information about accreditation and licensing. In the USA, for example, there is no registry of CLIA-licensed laboratories, nor is there any public record of laboratories' performance in proficiency testing or whether they participate in such proficiency testing.

RECENT DEVELOPMENTS IN REGULATION

So far, this chapter gives an overview of some regulatory frameworks for genetic tests and outlines some of the major regulatory gaps that are cause for concern. Policy discussions seem to be moving into a new phase of activity in regard to the regulation of genetic tests. In the USA, the past two years have seen an investigation of direct-to-consumer testing by the Government Accountability Office (US Government Accountability Office, 2006), the introduction of two bills on oversight of genetic testing into the US Senate (US Congress, 2007a, 2007b) and the Secretary's Advisory Committee on Genetics, Health and Society (SACGHS, 2008) has published a new report that analyses the regulatory framework and recommends a wide range of changes.

Meanwhile, in the UK, the Human Genetics Commission (HGC) has published a follow-up to its 2003 report on the regulation of direct-to-consumer (DTC) genetic tests, intended to prompt the government to act (Human Genetics Commission, 2007). The HGC's recommendations included preventing DTC provision of predictive tests, ensuring pre-market review of new tests and establishing a code of practice to govern DTC testing services. At the European level, consultation has begun on revising the medical device directives including the IVD Directive, with much expectation that the system will become more prescriptive and plug some of the current regulatory gaps, in part to address concerns about genetic tests (European Commission, 2007). The Dutch government has issued a report outlining a new model for risk classification in Europe that would see the reclassification of genetic tests as moderate risk (Hollestelle and de Bruin, 2006).

Another European body, the Council of Europe, is also driving policy in this area. Its 2008 Protocol on Genetic Testing establishes a general rule that a 'genetic test for health purposes may only be performed under individualised medical supervision' (Council of Europe, 2008). The impact of the protocol will depend on how many states choose formally to ratify it and thereby accept it as a legally binding protocol.

Finally, in Australia the Therapeutic Goods Administration (TGA), the body responsible for licensing IVD devices, has revised its regulations partly in a response to concerns expressed by successive Australian governments about the regulation of genetic tests. The TGA intends to implement regulatory mechanisms that will prohibit access to home use (self-testing) tests, including genetic tests, for serious disease markers. As part of this process the TGA has issued a guidance document about the regulation of nutrigenetic tests (Therapeutic Goods Administration, 2007).

In the EU, there have been considerable efforts to harmonize oversight of laboratory quality assurance systems through a number of national, regional and international schemes, culminating in the European Molecular Genetics Quality Network. Participants in the Network include 34 European countries and laboratories from Australia and the USA. These quality assurance initiatives have led to a new project—EuroGentest—an ambitious attempt to move beyond the previous focus on laboratory quality assurance to develop a series of discrete, but linked programs that deal with all aspects of quality in genetic testing services, from evaluation of the clinical validity and utility of tests to genetic counseling.

International activities are also significant, not least the new OECD guidelines on quality assurance for molecular genetics laboratories (OECD, 2007b). These examples all demonstrate that policy is already under development.

REGULATION OPTIONS

Policy development in this area is challenging because the desire to enhance the regulation of genetic tests to improve patient safety and ensure public confidence must be balanced against the need to promote innovation and provide timely access to tests. These conflicting principles are not the only concern as there is also the practical matter of resources. Any increase in regulation will place burdens on both test developers/providers and on regulatory agencies. This final section discusses some options for dealing with these challenges.

This analysis will draw on current trends in regulatory theory, in particular the concept of 'responsive regulation' developed by Ayres and Braithwaite (1992). Responsive regulation aspires to transcend traditional approaches to regulation based on a top-down system of regulation by the state, advocating instead a more de-centered and diffuse approach, which assumes that the state is not the only effective regulatory agent and that other actors can fulfill many key regulatory functions. A balance between the two approaches can be struck by creating systems where the state delegates core regulatory functions to other parties, acting as a 'meta-regulator' with standard-setting authority and ultimate powers of enforcement.

Similarly, traditional reliance on statutory controls enforced by draconian sanctions, often including use of the criminal law (Black, 2002; Scott, 2003) are rejected in the responsive regulation model, which favors an approach premised on minimum statutory regulation required to achieve desired outcomes. This is achieved by creating regulatory systems that are risk-based, i.e. proportionate to the dangers posed, and responsive, i.e. rewarding good conduct with more minimal intervention, but punishing bad practice with increasingly tough sanctions. This approach is captured in the concept of an enforcement pyramid.

The concept of regulation by information disclosure, an approach to regulation which focuses on addressing 'the asymmetries of information between consumers and traders' (Howells, 2005) is also drawn on. This strategy is now very popular in consumer protection fields because it is seen as a way of balancing the need to protect the public with a desire to encourage freedom of choice.

Pre-market review and post-market controls

The first issue to consider is how to develop an approach to pre-market review of new tests that is not unduly burdensome. Pre-market review can have a number of functions. For the most high-risk tests, regulators will use pre-market review to set out in detail the types of clinical studies they require a test developer to perform, as in the FDA's Pre-Market Approval process for Class III tests. Such an approach can require test developers to mount costly clinical trials and delay their entry to the market. There is uncertainty in this process about whether the test developer will be able to provide sufficient data to convince the regulator of the test's safety and effectiveness. For moderate-risk tests, a less burdensome approach to pre-market review can be taken, one which focuses on ensuring truth-in-labeling (as in the FDA's 510 k process for Class II devices). In this model, a more modest amount of data may be required to gain regulatory approval. Indeed, companies will often rely on the existing scientific literature to demonstrate the clinical validity of their tests, rather than mounting studies of their own. The focus here is on ensuring that any claims made for a test's performance are supported by scientific evidence; where there are problems, the regulator can moderate the company's claims rather than requiring them to mount fresh clinical studies.

The burden could further be controlled by limiting pre-market review to appraisal of data on analytic validity (how well the test identifies the gene/s) and clinical validity (the relationship between the gene/s and clinical status). This model would allow companies to reach market fast and to begin to attempt to recoup their investment. At the same time, it allows regulators to fulfill the most pressing obligation placed on them—addressing asymmetries of information by ensuring that companies do not make unsubstantiated and overblown claims for the tests they sell and that they provide test users with accurate and comprehensive information on their tests. However, this approach may allow tests onto the market with only limited evidence of their safety and effectiveness. Even where clinical studies have been mounted prior to market entry, routine clinical use of a new test is the only way to gain a full understanding of its safety and effectiveness. A least burdensome approach to pre-market review should be accompanied by more effective use of post-marketing controls. Similar arguments have been made about the merits of enhancing post-market rather than pre-market controls in drug regulation (Reed et al., 2006).

Test and laboratory registry

The model outlined here has much in common with one of the central recommendations of the Secretary's Advisory Committee on Genetic Testing (SACGT), whose initial report had recommended that where limited pre-market approval was given for a test, it should be accompanied by transparency of information (SACGT, 2000). Another approach that focuses on the concept of regulation by information disclosure has been suggested by some of the companies who are affected by the FDA's decision to regulate IVDMIAs. These companies are advocating an alternative system of regulation which would require LDT developers to provide details of their company, their laboratory and their tests on a web-based registry

(Radensky, 2008). This system would not entail submitting tests for pre-market review by FDA, but would provide some controls to ensure that companies provide clinical evidence to test users and provide FDA with the opportunity to comment on the quality of the evidence. Aspects of this proposal has now been taken up by SACGHS in its 2008 oversight report, which recommends a mandatory registry for all LDTs, although SACGHS does not state FDA should control the registry or that FDA's regulation of LDTs should be limited to a registry system (SACGHS, 2008).

Such a model is an intriguing one but it raises two important questions: why should LDT developers get an easier ride than kit manufacturers, and who deals with complaints? The latter problem could be addressed by the statutory licensing authority adopting a meta-regulator role, allowing the market to operate with minimal controls but ready to step in where concerns arise. Clearly, such a model has parallels with the EU's system, based as it is on reacting to complaints once a test is on the market. Its advantage over the secretive EU system, in which technical files are treated as commercially confidential, is that it introduces a high level of information disclosure. It carries certain risks for companies, as they would be prey to the use of complaints as a spoiling tactic by competitor companies and the threat that products might be recalled or withdrawn after they had entered the market. A minimal approach to pre-market review might provide a greater level of regulatory certainty.

Enhanced role for third parties

Part of the solution may be to delegate pre-market review to a third party, possibly in conjunction with more flexible standards or review processes tailored to the resources of small clinical laboratories. Use of third-party review can minimize the burden on regulators and allow them to concentrate their resources on the highest risk tests. In the UK, for example, completion of a gene dossier for the Genetic Testing Network, which advises the National Health Service on genetic testing, could be officially recognized as an acceptable alternative to pre-market review. In the USA, a similar status could be given to LDTs that have received pre-market review by New York State or have been through the Collaboration, Education and Test Translation (CETT) program review. The CETT program, developed by the US National Institute of Health Office of Rare Diseases, has a mandate to support translation of genetic research into clinical applications, reviews evidence regarding genetic tests for rare diseases and maintains a publicly accessible database of such tests.

This model would require cooperation on harmonized standard-setting and agreement on which tests pose sufficiently high risk that they require more stringent standards. Such a development can be seen in Australia's efforts to deal with LDTs. For low- and moderate-risk tests, laboratories must register with the Therapeutic Goods Administration and notify the agency about the tests they make. Test validation must meet TGA-endorsed standards, but will be carried out by the National Association of Testing Authorities and the National Pathology Accreditation Advisory Council, the bodies responsible for assuring laboratory performance in Australia. Only high-risk tests will be subject to the same standards and processes that apply to test kits but the TGA can investigate concerns about a test in lower risk categories.

Consumer education

Regulators can also take on educational initiatives. For instance, in July 2006, the Federal Trade Commission, in conjunction with the FDA and the Centers for Disease Control and Prevention, issued a consumer alert warning consumers to be wary of the claims made for DTC genetic tests (Federal Trade Commission, 2006). The US Government Accountability Office (GAO) at the same time criticized several nutrigenetic testing services it had reviewed (US Government Accountability Office, 2006). The Human Genetics Commission has called for the creation of an organization that can act as an independent, credible source of information on genetic tests to the general public (Human Genetics Commission, 2003).

CONCLUSION

Striking the appropriate balance between enhancing regulation and promoting innovation is difficult. There is little consensus even within stakeholder groups, let alone between them, about the way forward. Regulators are loath to overextend their limited resources, just as test developers rarely want to see an increase in their regulatory burden. Asking regulators to review more tests than they do now and asking more companies to submit a greater number of their tests for regulatory scrutiny places demands on both parties. What is required is a trade-off between depth of review and breadth of coverage. Focusing pre-market review on ensuring truth-in-labeling for all but the highest-risk tests and limiting that review to analytic and clinical validity is an attempt to strike such a balance. This chapter has outlined a model of regulation by information disclosure that seeks to redress this imbalance, tipping the informational scales back towards the test user, whether doctor, patient or health care insurer.

A range of solutions exists and it may be that one size does not fit all. A risk-based approach may be appropriate to triage the cases in greatest need of attention. For some tests it may be sufficient to rely on disclosure of data on a test registry without pre-market review (tests for genetic predisposition to restless legs syndrome, for instance, would seem to pose limited risks to patients). For others, such disclosure mechanisms may be best tied to pre-market review to ensure truth-in-labeling, while a smaller number of tests may require both heightened pre-market scrutiny and controls over who can order and receive test results. Deciding where nutrigenetic tests will fall within evolving regulatory systems will depend on a variety of factors including the scientific and clinical claims made for the tests and whether or not they are delivered direct-to-consumer.

References

Altenstetter, C. (2005). International collaboration on medical device regulation: issues, problems and stakeholders. In *Commercialization of health care: global and local dynamics and policy responses* (M. Mackintosh and M. Koivusalo, eds.). Palgrave, London, pp. 170–86.

Andersson, H.C., Krousel-Wood, M.A., Jackson, K.E., Rice, J. and Lubin, I.M. (2002). Medical genetic test reporting for cystic fibrosis (ΔF508) and factor V Leiden in North American laboratories. *Genet Med* 5:324–27.

Ayres, I. and Braithwaite, J. (1992). *Responsive regulation: transcending the deregulation debate.* Oxford University Press, New York.

Black, J. (2002). Critical reflections on regulation. London School of Economics, Carr discussion paper 4, available at: http://www.lse.ac.uk/collections/CARR/pdf/Disspaper4.pdf.

College of American Pathologists (2007). Pathology reporting. http://www.cap.org/apps/cap.portal?_nfpb=true&cntvwrPtlt_actionOverride=%2Fportlets%2FcontentViewer%2Fshow&_windowLabel=cntvwrPtlt&cntvwrPtlt%7BactionForm.contentReference%7D=pathology_reporting%2Fresource_site_desc.html&_state=maximized&_pageLabel=cntvwr.

Council of Europe (2008). Additional protocol to the Convention on Human Rights and Biomedicine, concerning genetic testing for health purposes, available online: http://conventions.coe.int/Treaty/EN/Treaties/Html/TestGen.htm.

European Commission (1993). Council Directive 93/42/EEC of 14 June 1993 concerning medical devices. European Commission, Brussels. Available online: http://www.translate.com/multilingual_standard/MDD.pdf.

European Commission (1998). Directive 98/79/EC of the European Parliament and of the Council of 27 October 1998 on in vitro diagnostic medical devices. European Commission, Brussels. Available online: http://eur-lex.europa.eu/LexUriServ/LexUriServ.do?uri=OJ:L:1998:331:0001:0037:EN:PDF.

European Commission (2004). Guidelines on post market clinical follow-up MEDDEV 2.12-2. European Commission, Brussels. Available online: http://ec.europa.eu/enterprise/medical_devices/meddev/2_12-2_05-2004.pdf.

European Commission (2007). Recast of the Medical Devices Directives – public consultation. European Commission, Brussels. Available online: http://ec.europa.eu/enterprise/medical_devices/consult_recast_2008_en.htm.

Evans, J.P., et al. (2001). The complexities of predictive genetic testing. *Br Med J* 322:1052–6.

FDA (2006a). Draft guidance for industry, clinical laboratories, and FDA staff – in vitro diagnostic multivariate index assays. CDRH, Bethesda.

FDA (2006b). Report of the postmarket transformation leadership team: strengthening FDA's postmarket program for medical devices. FDA, Washington, DC.

Federal Trade Commission (2006). At home genetic tests: a healthy dose of skepticism may be the best prescription. FTC, Washington, DC.

Foucar, E. (2001). Predictive genetics and predictive morphology have certain similarities. *Br Med J* 323:514.

GeneWatch UK (2002). Genetic testing on the High Street. http://www.genewatch.org/uploads/f03c6d66a9b354535738483c1c3d49e4/SocionaBrief.rtf.

Godard, B., et al. (2003). Provision of genetic services in Europe: current practices and issues. *Eur J Hum Genet* 11(Suppl. 2):13–48.

Haga, S., et al. (2003). Genomic profiling to promote a healthy lifestyle: not ready for prime time. *Nat Genet* 34:347–50.

Harris, R. and Reid, M. (1997). Medical genetic services in 31 countries: an overview. *Eur J Hum Genet* 5(Suppl 2):3–21.

Hogarth, S. (2007). The clinical application of new molecular diagnostic technologies – a review of the regulatory and policy issues, a report to Health Canada.(Unpublished).

Hollestelle, M. and de Bruijn, A. (2006). In vitro diagnostic medical devices. Decision rules for ivd-classification. National Institute for Public Health and the Environment, Bilthoven. Available online: http://www.rivm.nl/bibliotheek/rapporten/360050007.pdf.

Howells, G. (2005). The potential and limits of consumer empowerment by information. *J Law Soc* 32:349–70.

Hudson, K., et al. (2006). Oversight of US genetic testing laboratories. *Nat Biotechnol* 24:1083–90.

Human Genetics Commission (2002). Minutes of Genetic Services Sub-group, 19 April 2002. Available online: http://www.hgc.gov.uk/Client/Content_wide.asp?ContentId=686.

Human Genetics Commission (2003). Genes direct – ensuring the effective oversight of genetic tests supplied directly to the public. Department of Health, London.

Human Genetics Commission (2007). More genes direct: a report on developments in the availability, marketing and regulation of genetic tests supplied direct to the public. Department of Health, London.

Ibaretta, D., Bock, A., Klein, C. and Rodriguez-Cerezo, E. (2003). Towards quality assurance and harmonisation of genetic testing services in the EU. Institute for Prospective Technological Studies and European Commission Joint Research Centre, Brussels.

Khoury, M.J., Little, J. and Burke, W. (eds.) (2004). *Human genome epidemiology: a scientific foundation for using genetic information to improve health and prevent disease*. Oxford University Press, Oxford, p. 7.

Krousel-Wood, M., Andersson, H.C., Rice, J., Jackson, K.E., Rosner, E.R. and Lubin, I.M. (2003). Physicians' perceived usefulness of and satisfaction with test reports for cystic fibrosis (Δf508) and factor V Leiden. *Genet Med* 5:166–67.

New York State (2007). Clinical laboratory evaluation program – submission guidelines for test approval. NY State, New York.

OECD (2007a). Genetic testing: a survey of quality assurance and proficiency standards. OECD, Paris.

OECD (2007b). Guidelines on quality assurance in molecular genetic testing. OECD, Paris.

Petit, E., Tasse, A.M. and Godard, B. (2008). An empirical analysis of the legal frameworks governing genetic services labs in Canadian provinces. *Hlth Law Rev* 16:65–72.

Pilot, L. (1999). Medical device labeling in the European Union. *Med Device Diagnos Ind* May.

Radensky, P. (2008). Comments to SAGGHS in Minutes of meeting of 12 February 2008. http://www4.od.nih.gov/oba/SACGHS/meetings/2008Feb/transcript/fulldayFeb12.pdf.

Reed, S., Califf, R.R. and Schulman, K. (2006). How changes in drug-safety regulations affect the way drug and biotech companies invest in innovation. *Hlth Affairs* 25:5.

SACGHS (2008). US system of oversight of genetic testing. NIH, Bethesda.

SACGT (2000). Enhancing the oversight of genetic tests: recommendations of the SACGT Secretary's Advisory Committee on genetic testing. NIH, Bethesda.

Scott, C. (2003). Regulation in the age of governance: The rise of the post-regulatory state. National Europe Centre, Paper No.100, available online at: http://www.anu.edu.au/NEC/NEC%20EVENTS/Events%202003/scott1.pdf.

Therapeutic Goods Administration (2007). The regulation of nutrigenetic tests in Australia: guidance document. http://www.tga.gov.au/devices/ivd-nutrigenetic.htm.
US Congress (1994). Federal Food, Drug and Cosmetic Act, 21 U.S.C. § 360i (a).
US Congress (2006). At home DNA tests: marketing scam or medical breakthrough: hearing before the s. spec. comm. on aging. 109th Cong. pp. 109–707.
US Congress (2007a). Genomics and personalized medicine Act. S.976, 110th Congress.
US Congress (2007b). Laboratory test improvement Act, S.736, 110th Congress.
US Government Accountability Office (2006). Nutrigenetic tests: tests purchased from four websites mislead consumers. GAO, Washington, DC.
Willey, A. (2007). New York State's clinical laboratory evaluation programme – presentation to SACGHS March 2007 http://www4.od.nih.gov/oba/SACGHS/meetings/Mar2007/Mon%20pm%20-%20Willey.pdf.
Winn-Deen, E.S. (2003). Fulfilling the promise of personalized medicine. *IVD Technol.* November/December.

Further reading

Gunningham, N. and Grabosky, P. (1998). *Smart regulation: designing environmental policy.* Oxford University Press, Oxford.

CHAPTER

5

Risk-Based Regulation of Direct-to-Consumer Nutrigenetic Tests

Nola M. Ries

OUTLINE

Nutrition and Genomics
ISBN: 978-0-12-374125-7

SUMMARY

An increasingly wide variety of genetics tests are marketed and sold directly to consumers (DTC) via the Internet, including nutrigenetic tests. The initial entry of DTC genetic tests in the marketplace generated calls for bans or restricted access only through medical professionals. The growing range of DTC tests now available has prompted advocacy of risk-based regulation. This chapter discusses risks and potential benefits claimed to be associated with DTC genetic tests and observes that more research about the nature and magnitude of risks is needed to inform appropriate regulation. The chapter considers where nutrigenetic tests fall into current attempts to categorize such tests and notes that they straddle a boundary between 'medical' and 'lifestyle' tests. The chapter examines various regulatory tools that may be applied to DTC genetic tests and favors improved information disclosure about the strengths and limits of the current generation of DTC genetic tests. It also endorses efforts to use existing consumer protection laws to police companies that make false or misleading claims.

INTRODUCTION

This chapter discusses direct-to-consumer (DTC) advertising and sale of genetic tests, specifically nutrigenetic tests, that claim to identify genetic variations thought to affect nutrient metabolism and predispose individuals to diseases risks that may be mitigated through dietary and other lifestyle modifications. A wide range of genetic tests is currently marketed online to consumers, including paternity testing, genealogical testing, testing for health purposes, tests that purport to identify behavioral traits and tests claiming to help find a physiologically compatible love match. Genetic tests available directly to consumers are variously described as falling into 'medical', 'lifestyle', 'enhancement' and 'recreational' categories and raise differing regulatory concerns. Nutrigenetic tests presently straddle a boundary between these categories. As Paula Saukko's chapter notes (Chapter 11), companies that sell DTC nutrigenetic tests adopt divergent marketing schemes, some portraying tests as medical, others promoting the tests as lifestyle enhancements.

Early days of DTC marketing of genetic tests prompted calls to restrict DTC sales and require that tests only be provided through a health care provider. Several realities challenge that position. First, not all genetic tests are alike and a blanket prohibition on DTC provision of any genetic test may be excessive, both in imposing undue enforcement burdens on regulators and denying consumers access to tests that may be useful or, at least, mostly harmless. Second, requiring that a health care provider is involved in genetic testing places additional demands on professional capacity, as discussed in other chapters. Third, legislative bans may fail to meet consumer protection goals since the online marketplace allows consumers to access services and products from companies operating in other countries that are beyond the jurisdiction of domestic prohibitions on DTC tests. The borderless world of online commerce demands other regulatory tools, especially those that aim to educate consumers so they may make informed purchasing decisions.

This chapter begins with an overview of types of health-related genetic testing services currently advertised and sold DTC. It summarizes harms and benefits said to be associated with DTC genetic tests, then examines options for regulating DTC genetic tests, including prohibitions on DTC tests, allowing DTC access only for tests considered to be of lower risk, enforcing truth-in-advertising legislation against DTC companies that make false and/or misleading claims and publicizing educational resources to help consumers make more informed decisions about purchasing genetic tests. The chapter concludes that more nuanced consideration must be given to categorizing genetic tests, evaluating their harms and benefits and the relevance of these points to regulatory measures. Regulation should be commensurate with risk, but clearer evidence of risks of DTC access to genetic tests is needed to inform balanced and fair rules. As the American Society of Human Genetics (2007) observes, 'a one-size-fits-all [regulatory] approach is not appropriate for DTC tests, because the types of tests being offered are heterogeneous, and their consequences are wide ranging'. Given the state of flux in regulatory debates about DTC genetic tests, the steady emergence of research results (especially from genome-wide association studies) and new entrants to the DTC genetic testing industry, a useful, immediate step is to coordinate efforts in establishing and promoting a publicly accessible database of genetic test information. This will help consumers, health care professionals and other interested parties understand the current state of knowledge about genetic tests, make decisions about using such tests and interpret the meaning of test results.

TYPES OF DTC GENETIC TESTS ON THE MARKET

A growing number of commercial firms are advertising and selling genetic tests directly to consumers. A February 2008 article in *Nature Biotechnology* observes:

> The onset of genetic testing as a wide-spread consumer commodity continues to gather pace. At least 27 web-based companies now offer genetic tests – once the exclusive domain of hospital clinics and academic laboratories – directly to consumers for costs ranging from roughly $100 for a simple gene scan to $350 000 for a personal genome sequence and related medical advice (Schmidt, 2008).

In a December 2007 report, the UK Human Genetics Commission (2007, Appendix 2) identified 26 companies based in the USA, UK and Europe that advertise and/or sell DTC genetic tests. While not an exhaustive count of the number of firms currently operating in this field, the tests provided by these 26 companies reveal the range of health concerns for which testing is available and the variable models by which services are offered. DTC tests assess genetic predisposition for a range of diseases and conditions, including cancer (mainly breast, ovarian, prostate and colorectal cancer), cardiovascular disease, osteoporosis, diabetes (type 1 and type 2), obesity, celiac disease, inflammatory bowel disease, hemochromatosis, factor V Leiden, thrombophilia, glaucoma, macular degeneration and Alzheimer's disease. They may test for genetic factors that affect lipid, glucose, caffeine, alcohol and pharmaceutical metabolism. Pregnancy and newborn screening tests are also available. Some tests are bundled into packages marketed to men (e.g. including tests for genetic predisposition for prostate cancer and cardiovascular disease) and women (e.g. including tests for genetic predisposition to breast cancer and osteoporosis). A nutrigenetic test package may test for various genes that affect nutrient metabolism and susceptibility to common diseases that may be mitigated through dietary changes. Some tests aimed at fitness enthusiasts or athletes offer information to help optimize physical training and nutrition to enhance performance.

Genetic tests that are sold DTC require a consumer to order the company's test kit, collect a genetic sample at home (generally a buccal swab), return it to the company for analysis and, for some tests, provide lifestyle information, including information about sex, age, health history, diet and other behaviors (e.g. tobacco use, exercise). Exchange of personal information and test results may be done online or by mail and counseling, where provided, may be done by telephone or by referral to in-person counseling. Business models vary among genetic testing companies. Some provide only one or a few types of tests, while others offer a broader range. Some sell tests directly to consumers, while others simply advertise DTC but only accept testing referrals from a health care professional (typically a physician). Some provide post-test counseling with a health care professional (such as a dietician, genetic counselor or physician) to explain test results and discuss potential lifestyle modifications or need for medical follow-up. If counseling is offered, it may be included in a standard fee or it may be an optional service available at extra cost. In addition to genetic testing services, some companies sell nutritional supplements claimed to be formulated to improve heart health, bone health or other problem areas identified through the company's genetic testing. In some cases, companies market packages that include genetic testing, consultation and, sometimes, diet plans and supplements, aimed at mitigating disease susceptibility or achieving weight loss.

In an analysis of 24 DTC genetic testing companies, Geransar and Einsiedel (2008) examined the companies' target markets (tests marketed to consumers only or health professionals and consumers both), company policies regarding involvement of physicians in the testing process (the consumer's physician or a physician associated with the company, or no involvement) and company policies regarding genetic counseling (counseling in person, via telephone or other means, consumer's physician responsible for referral to genetic counselor, referral to contracted counselor, or no requirement for genetic counseling). They also examined the types of information on company websites about the genetic condition(s) for which tests are sold (e.g. information about disease symptoms, prevention, treatment) and cited sources of information (e.g. peer-reviewed studies, unpublished company studies, professional organizations). Companies that sell nutrigenetic tests were less likely to require physician involvement (compared, for example, to companies that sell diagnostic or predisposition tests for conditions like cystic fibrosis, Huntington disease, Alzheimer's disease and breast and ovarian cancer) and to offer counseling by telephone or other long-distance means.

POTENTIAL HARMS AND BENEFITS OF DTC GENETIC TESTS

Potential harms and benefits of DTC genetic tests relate to four areas:

1. the test results themselves
2. consumers' emotional reaction to the test results
3. consumers' behavioral reaction to the test results
4. broader social issues.

These are summarized in Figure 5.1.

Genetic tests do not have uniform risks or benefits and speculation about positive and negative effects of learning new genetic information is not always borne out in reality. Organizations like the UK Human Genetics Commission (2003, 2007) and the US Secretary's Advisory Committee on Genetics, Health and Society (2007, 2008) have distinguished between 'medical' and 'lifestyle' and 'higher-risk' and 'lower-risk' tests, but acknowledge these distinctions are not precise. Nutrigenetic tests are often referred to as lifestyle (Human Genetics Commission, 2007) or 'enhancement' tests (e.g. Geransar and Einsiedel, 2008), suggesting a non-serious or frivolous application relevant only to those preoccupied with health, fitness and being an early technology adopter. But genetic tests that indicate an elevated risk of diabetes, cardiovascular disease or other conditions that can be modified by dietary changes are not trivial.

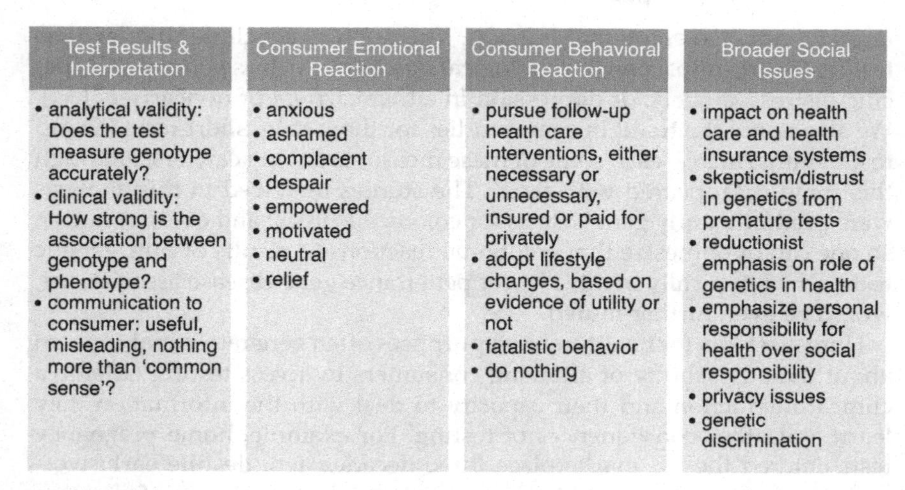

Test Results & Interpretation
• analytical validity: Does the test measure genotype accurately?
• clinical validity: How strong is the association between genotype and phenotype?
• communication to consumer: useful, misleading, nothing more than 'common sense'?
Consumer Emotional Reaction
• anxious
• ashamed
• complacent
• despair
• empowered
• motivated
• neutral
• relief
Consumer Behavioral Reaction
• pursue follow-up health care interventions, either necessary or unnecessary, insured or paid for privately
• adopt lifestyle changes, based on evidence of utility or not
• fatalistic behavior
• do nothing
Broader Social Issues
• impact on health care and health insurance systems
• skepticism/distrust in genetics from premature tests
• reductionist emphasis on role of genetics in health
• emphasize personal responsibility for health over social responsibility
• privacy issues
• genetic discrimination

FIGURE 5.1 Potential harms and benefits of DTC genetic tests.

At present, however, genetic tests vary widely in the specificity and certainty of information they provide and their utility in informing treatment options or behavior modifications. Burke et al. (2001) describe four categories that apply to health-related genetic tests:

1. tests with high predictive value where effective treatment is available
2. tests with high predictive value where no effective treatment is currently available
3. tests with low predictive value where no effective treatment is available
4. tests of low predictive value but effective preventive measures exist.

Each one of these categories of tests presents different potential harms and benefits.

In the first category, benefits of testing are clearly high; individuals who find out about a genetic predisposition before disease onset may take measures to reduce development or severity of disease. Where a test is highly predictive and treatment is available, harm may arise from the treatment (e.g. if it is highly invasive or costly) and from personal and social reactions to one's status as at-risk for disease. The extent of these psychosocial harms will depend on the nature of the disease, the prevailing social context (i.e. the extent to which the condition is considered embarrassing or stigmatizing) and one's own emotional resiliency. These latter psychosocial harms are the key risks in situations where a test is highly predictive but no treatment is available.

In a systematic review of 65 studies examining emotional and behavioral impacts of genetic testing, Heshka and colleagues (2008) found little evidence that learning of an increased genetic risk for disease leads to ongoing adverse emotional impacts. The authors conclude that 'genetic testing had no impact on psychological outcomes such as general and specific distress, anxiety, or depression in either carriers or noncarriers [...]. We also noted the trend in some studies for there to be short-term (i.e. up to 4 months) increases in some of these measures among carriers, although this trend disappeared with time'. The studies reviewed in this analysis were predominantly genetic tests for colorectal, breast and ovarian cancer, so one can hypothesize that emotional reactions to results of nutrigenetic tests, which typically involve lower penetrance gene-disease associations, would be even further muted.

Home test kits for health-related purposes often generate initial concern about the advisability of allowing consumers to access testing outside a clinical interaction and their capacity to deal with the information they learn and other consequences of testing. For example, home pregnancy tests entered the US marketplace three decades ago, despite early worries about accuracy of test kits and allowing women to learn information about their pregnancy status outside a clinical setting. A letter writer in the *American Journal of Public Health* expressed the view in 1976 that, 'I am becoming increasingly worried by the escalating use of such kits by nontechnical staff ... and especially, as in this case [of home pregnancy tests], by the patient herself' (Entwistle, 1976). Today, home pregnancy tests are widely accepted and used. Home HIV test kits have raised similar concerns. Opponents of home HIV testing emphasized psychological hazards of receiving a positive test result in the absence of immediate counseling, impacts on interpersonal relationships, user errors and harms of false test results. It was predicted that testing uptake would largely be among the affluent and 'worried well' (for discussion, see e.g. Walensky and Paltiel, 2006). Proponents pointed to potential for broader testing, particularly for people who did not wish to attend a clinic and knowledge of HIV status to help limit spread of infection (for further discussion, see e.g. Fabbri, 1995). The US FDA banned HIV home testing kits in the late 1980s but approved a home collection test in 1996 (Food and Drug Administration, 2008).

Tests with high predictive value underscore concerns about emotional and behavioral reactions to relatively certain information. Tests with low predictive value, however, raise different concerns. If predictive accuracy is uncertain, but effective treatment is available, Burke and colleagues say the benefit of testing may outweigh harm, especially where the treatment is not especially burdensome. Better to undergo treatment for a disease of uncertain risk of occurrence than to forgo treatment and end up with a disease that could have been prevented or mitigated. Where a genetic test has low predictive value and no effective treatment is available, the primary harm may be having spent money for nothing.

Some nutrigenetic tests currently sold DTC are criticized as having low predictive value because the strength of associations between particular genetic variants and disease risk are not clear. In a study of DTC genetic tests sold by seven companies that purport to give personalized nutrition and lifestyle advice, Janssens and colleagues (2008) conclude:

> Our review of meta-analyses found significant associations with disease risk for fewer than half of the 56 genes that are tested in commercially available genomic profiles. Various polymorphisms of these genes were associated with risk for 28 different disorders. Many of these disorders were unrelated to the ostensible target condition, and the associations were generally modest.

They go on to predict: 'Despite advances in nutrigenomics and pharmacogenomics research, it could take years, if not decades, before lifestyle and medical interventions can be responsibly and effectively tailored to individual genomic profiles' (Janssens et al., 2008). Some nutrigenetic test results offer advice that is relatively generic, as Saukko describes in Chapter 11, and some assert that nutrigenetic tests mislead consumers by claiming to provide 'personalized' nutrition advice that simply reflects healthy eating advice applicable to everyone (see e.g. Government Accountability Office, 2006).

Primary concerns with DTC sale of genetic tests for conditions like cancer and Alzheimer's disease (or non-genetic tests for conditions like HIV) are that consumers will have harmful emotional and behavioral reactions. In contrast, the primary concern with DTC nutrigenetic tests is that the evidence base for some claims is presently weak and consumers are paying money for generic advice wrapped up in claims of personalization. Different consumer protection issues arise for different types of DTC genetic tests and a challenge for regulators is to sort out where to allocate law-making and enforcement resources.

REGULATORY INITIATIVES REGARDING DTC ADVERTISING AND SALE OF GENETIC TESTS

Regulatory options for DTC genetic tests, as summarized in Figure 5.2, include outright prohibitions on direct sale to consumers, rules to permit DTC access only for tests considered to be of lower risk, enforcement of existing 'truth-in-advertising' legislation or enhancing information provision to help consumers make more knowledgeable decisions about purchasing genetic tests. Regulators may also adopt a *laissez-faire* approach that leaves a marketplace largely unregulated (or even if regulations might apply, they are not enforced).

Prohibitions on DTC Genetic Tests

DTC genetic testing is prohibited in some jurisdictions. In the USA, a 2007 review found that 13 states prohibit DTC testing and '26 states and

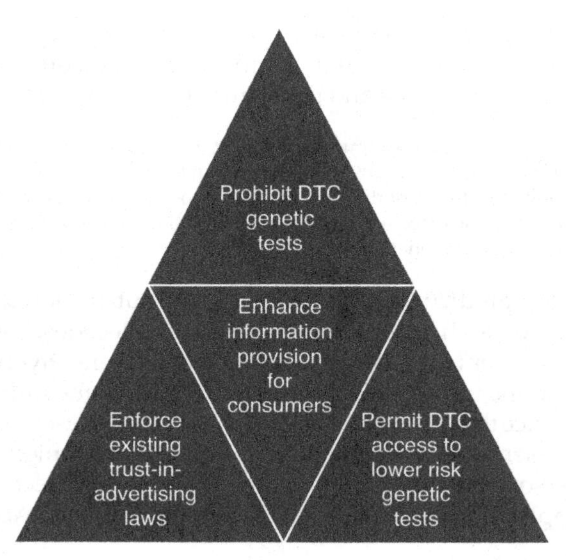

FIGURE 5.2 Regulatory options for DTC genetic tests.

the District of Columbia permit DTC laboratory testing without restriction ... while 11 permit it only for specified categories of tests, which tend to exclude genetic tests' (Genetics and Public Policy Center, 2007). In 2008, the States of New York and California issued cease-and-desist letters to a number of genetic testing companies asserting they violated state law by selling DTC tests in the absence of a physician's order (Wadman, 2008). The Council of Europe 2008 Protocol on Genetic Testing states that a 'genetic test for health purposes may only be performed under individualized medical supervision' (Council of Europe, 2008), though so-called 'lifestyle' tests are generally considered exempt from European regulations governing genetic tests (Melzer et al., 2008). Some health professional organizations also advocate that genetic tests should only be undertaken with involvement of a qualified and knowledgeable professional (see e.g. American College of Medical Genetics, 2004, 2008; American Medical Association, 2008).

While DTC access is prohibited outright in some jurisdictions or permissible only with physician involvement in others, the usefulness and feasibility of restrictions are increasingly questioned, especially for tests with lower risks where at least some consumers seek direct access. The UK Human Genetics Commission commented in its 2007 report, *More Genes Direct*:

> ... we are now seeing a burgeoning cottage industry in so-called 'lifestyle' tests together with the regimens, dietary supplements and self-administered medications

that they are claimed to indicate. Some of these tests are relatively innocuous. There may even be benefits in that they provide reassurance or a sense of empowerment, or encourage the adoption of a healthy exercise regime (Human Genetics Commission, 2007).

In a 2007 draft report, the US Secretary's Advisory Committee opined 'that applying the same regulatory framework to every genetic test is infeasible given the number of tests in use and in development and the costs and resources that would be needed to support such a structure. Moreover, such a policy could unnecessarily delay patient access to important new technologies' (Secretary's Advisory Committee on Genetics, Health and Society, 2007).

Leaving the market largely unfettered is one response to 'relatively innocuous' services and products but the problem, as the Human Genetic Commission (2007, p. 3) points out, is that:

> other tests are available with potentially more serious implications. Tests that claim to predict the onset of disease or indicate a heightened risk of serious conditions, or, alternatively, to offer peace of mind and the promise of a long and active retirement, can significantly influence choices that profoundly and enduringly affect an individual's health.

In the context of DTC genetic tests, then, regulators attempt to weigh the potential harmful impacts of consumers receiving health-related information outside the context of a traditional clinical interaction. This leads to an approach to prohibit or restrict direct provision of more serious (or riskier) tests, while allowing less serious (or lower risk) tests to be marketed and sold directly to consumers.

Permitting DTC Access to Lower-Risk Genetic Tests

An increasingly common suggestion is to regulate genetic tests in proportion to their risks. This intuitively sensible approach to regulation is not new and many regulatory frameworks assess products based on risk level, such as medical device regulation. The tricky aspect of this regulatory approach is in determining the nature and magnitude of risks associated with a particular activity. Regulation is easier when potential benefits and harms can be measured with some specificity and objectivity and more complex where there is debate and lack of evidence about harms and benefits.

Risk-based regulation has been advocated for regulation of DTC genetic tests. In a 2003 report on DTC genetic testing, for instance, the UK Human Genetics Commission advocated stricter controls on DTC genetic tests but suggested, in analogy to pharmaceuticals, that differing levels of control are appropriate for different types of genetic tests. The Commission argued at that time that: 'there is support for considering most genetic tests as if they were "prescription-only"', but attenuated that view 4 years later, acknowledging the growth in genetic testing services, including those in the 'relatively innocuous' category.

Assessing risks and possible benefits of DTC access requires speculation about emotional and behavioral reactions of consumers to genetic information and consequent social implications. The difficulty in predicting reactions to results of home tests has prompted some to argue that regulators should concern themselves only with regulating compliance of DTC providers with appropriate standards of test accuracy (see Chapter 4). In the context of home HIV test kits, for example, Salbu (1994) contended that weighing 'potential emotional reactions to reliable information ...' is outside appropriate regulatory purview and that regulators '... should consider only evidence that sheds light on whether the product is made so that consumers can use it without unreasonable risk of injury from the product itself and whether the product achieves its purported purposes within reasonable effectiveness parameters' (Salbu, 1994).

For DTC genetic tests, the sample collection is unlikely to cause injury, so the question is whether the test achieves its purposes. For nutrigenetic tests, this raises the blurry line between medical and lifestyle/enhancement tests. In a risk-based regulatory scheme where medical tests face more stringent scrutiny, companies may have an incentive to characterize their tests as offering lifestyle-oriented information that does not fall into a traditional medical paradigm. Indeed, some DTC genetic companies that received cease-and-desist orders from the States of New York and California in 2008 responded by arguing that they do not provide medical testing, but rather are 'enabling consumer access to research knowledge' (Avey, 2008), especially information emerging from genome-wide association studies.

To provide appropriate risk-based regulation of genetic tests, stakeholders will need to give more thorough consideration to the risks that are relevant and most likely to arise from DTC provision. Most debate about DTC genetic tests assumes regulators ought to consider how individuals react to and use information, but this assumption, too, might warrant debate. An editorial in *Nature Genetics* about DTC genetic tests for complex, common disorders queries 'whether this information will make a difference in their lives' (Editorial, 2007). If an individual undergoes nutrigenomic testing, will they actually adopt dietary modifications to lead healthier lives? As Bouwman and van Woerkum discuss in Chapter 7, numerous factors influence eating behavior and Caulfield and colleagues (see Chapter 12) point out structural barriers to healthy food access. But requiring genetic testing to make a difference in individual lives imposes a higher standard on it than many other services or products. Moreover, consumer reactions to and use of genetic information will vary and the actual influence of genetic information on personal behavior and attitudes may be difficult to measure, both in the short and long term.

Some research currently underway will help elucidate consumer reactions to and use of genetic information. The US Coriell Institute for Medical Research, a non-profit medical research group, offers free genetic testing to an anticipated 10 000 volunteers through the Corriel Personalized Medicine Collaborative (see http://www.coriell.org/index.php/content/view/92/167/). Participants will receive test results and advice about lifestyle modifications to reduce disease risks. As another example, Navigenics, a US-based DTC genetic testing company, is co-sponsoring clinical trials with researchers at the Mayo Clinic to 'examine how participants react psychologically and behaviourally to medical risk information based on different sources, including family history and genetic testing, and presented to them with or without counselling' (Pearson, 2008). These types of studies will improve understanding of individual responses to genetic information and the most effective means to communicate risk probabilities and counseling regarding recommended lifestyle changes.

Risk-based regulation of genetic tests is a step in the right direction yet likely a difficult area in which to reach stakeholder agreement. Nearly a decade ago, the US Secretary's Advisory Committee on Genetic Testing (2000) raised the problem of categorizing genetic tests based on assessment of risks and benefits. After much consultation and debate, the Committee concluded in a 2001 report on test classification that:

> fundamental, irresolvable questions had been raised about the feasibility of categorizing tests for oversight purposes based on a limited set of elements in a simple, linear fashion. Thus, the Committee decided that further efforts to develop a classification methodology for genetic tests should be curtailed for the present (Secretary's Advisory Committee on Genetic Testing, 2001).

The wide variety of DTC genetic tests now available likely exacerbates the challenge of classifying genetic tests, but regulation must take account of the differences among tests to achieve a balance between consumer protection and autonomy goals, as well as to provide fair rules for companies.

The problems of categorizing genetic tests, when there is much flux in the DTC genetic testing industry and still a lack of data about consumer reactions to information obtained from testing, may stall regulatory restrictions on DTC access. Nonetheless, other tools can be used to protect consumers from unscrupulous companies who make false or misleading claims and to arm consumers with information to make informed purchasing choices.

Enforcement of Truth-in-Advertising Laws

Laws regulating advertising of goods and services may regulate advertising of DTC genetic tests. In the USA, for example, the Food and Drug Administration may regulate false or misleading labeling of devices and the Federal Trade Commission (FTC) has enforcement authority for false, unfair and deceptive commercial practices (Ries, 2008). The FTC's law enforcement and consumer education program, 'Operation Cure.all', scouts out unproven health-related products and services marketed and sold on the Internet. In the past, the FTC, other US federal and state bodies, consumer protection organizations, along with North American and international partners, conducted two 'Health Claim Surf Days' to identify deceptive Internet advertising for health products. The FTC reports that:

> The surf days identified approximately 800 World Wide Web sites and numerous Usenet newsgroups that contain questionable promotions for products or services purporting to help cure, treat or prevent six diseases: heart disease, cancer, AIDS, diabetes, arthritis, and multiple sclerosis. After each Surf Day, web sites were sent e-mail messages, alerting them that their claims require scientific substantiation and that disseminating false or unsubstantiated claims violates federal law. Following the e-mail, FTC staff surveyed a representative sample ... and found that 28 percent of the sites had either removed their claims or had been taken down. [The Director of the FTC's Bureau of Consumer Protection said] that the agency is very encouraged by the fact that over a hundred sites making questionable claims voluntarily cleaned up their act, but warned that the agency will continue to monitor the Web for fraud and deception and bring law enforcement cases as appropriate (Federal Trade Commission, 1999).

While this program provides a useful precedent for policing claims, neither the FDA nor the FTC has focused enforcement resources on DTC genetic tests (American Society of Human Genetics, 2007). This is due, in part, to fragmented regulatory jurisdiction between the FDA and FTC (Ries, 2008). The FDA has warned some dietary supplement manufacturers about making impermissible claims that their products can treat, cure or prevent disease. Companies that engage in tied selling of nutrigenetic tests and dietary supplements may face penalties in relation to product claims. For example, a US government report on nutrigenetic testing described a DTC testing company that marketed Cat's Claw as a 'DNA repair' supplement, despite the lack of peer-reviewed evidence substantiating the plant's pharmacological effects. The report notes that '[t]he FDA has already sent Warning Letters to several dietary supplement manufacturers who explicitly claimed that Cat's Claw could help treat cancer and arthritis' (Government Accountability Office, 2006) and similar action could be taken against the DTC genetic testing company if the DNA repair claim violates federal rules.

In the UK, the Human Genetics Commission and GeneWatch have both pursued complaints about claims made in advertising for (alleged) genetic services and products. Hogarth et al. (2005) report that the Commission complained to the UK Advertising Standards Authority in 2003 about an advertisement about a 'Genetic Hair' product that falsely implied it used genetic technology to treat hair loss. The Advertising Authority upheld the complaint. In 2004, GeneWatch complained about misleading claims for NicoTest, a personalized testing service that claimed to help people stop smoking. The governmental authority, the Trading Standards Office, accepted the complaint, but the company revised its website claims before further action was taken.

These examples illustrate the use of truthful advertising rules to crack down on unsubstantiated claims in DTC health-related products and services. Regulatory agencies can improve the effectiveness of these tools by educating companies about permissible claims. For instance, the American Society of Human Genetics (2007) statement on DTC genetic tests advocates that: '[t]he FDA and the FTC should work together to develop guidelines for DTC testing companies to follow, to ensure that their claims are truthful and not misleading and that they adequately convey the scientific limitations for particular tests'. Providing such guidance will help companies in the DTC genetic testing industry understand the rules and facilitate voluntary compliance.

Enhancing Information Disclosure

Restrictions or prohibitions on DTC advertising and sale of genetic tests seek to protect consumers in a context where there is knowledge imbalance between the genetic testing company and the consumer. Those who are concerned about DTC sale of genetic tests worry that companies may take advantage of uninformed or gullible consumers who do not fully understand the nature and limits of the test and any associated products, such as personalized dietary advice and supplements. Regulatory and consumer protection bodies in some jurisdictions have issued public communication statements that send a 'buyer beware' message about DTC genetic tests. For example, the US Federal Trade Commission (2006) gives the following advice to consumers who are considering DTC genetic tests:

> Be wary of claims about the benefits these products supposedly offer. Some companies claim that at-home genetic tests can measure the risk of developing a particular disease, like heart disease, diabetes, cancer, or Alzheimer's. But the FDA and CDC say they aren't aware of any valid studies that prove these tests give accurate results. Having a particular gene doesn't necessarily mean that a disease will develop; not having a particular gene doesn't necessarily mean that the disease will not.
>
> Some companies also may claim that a person can protect against serious disease by choosing special foods and nutritional supplements. Consequently, the results of their at-home tests often include dietary advice and sales offers for 'customized' dietary supplements. But the advice rarely goes beyond standard sensible dietary recommendations. The FDA and CDC say they know of no valid scientific studies showing that genetic tests can be used safely or effectively to recommend nutritional choices.

Yet, the science base for genetic tests is changing as new research results emerge. Instead of blanket warnings against DTC genetic tests, some organizations now advocate a publicly available genetic test database or registry to provide consumers, health care professionals and other interested people with the current state of scientific knowledge about specific gene variants and genetic tests. The Secretary's Advisory Committee (2008) notes that:

> [t]here is currently no requirement that test providers disclose information to support claims about the accuracy and validity of testing and no central or uniform mechanism for providing this information in an accessible format to patients and providers.

The Committee recommends 'establishment of a voluntary system of genetic test registration through a public-private partnership', such as expanding the National Institutes of Health-funded GeneTests initiative, which provides 'current, authoritative information on genetic testing and its use in diagnosis, management, and genetic counseling ...' (see www.genetests.com).

In the UK, the Public Health Genetics Foundation and Royal College of Pathologists has recommended stronger oversight of claims made by diagnostic laboratory tests, including DTC genetic tests (Furness et al., 2008). They recommend a publicly accessible test database that includes information on a test's analytic validity, clinical validity and clinical utility, including information about gaps in that evidence. The vice-president of the Royal College of Pathologists reportedly emphasized the benefit of stricter scrutiny of claims: 'Then companies would have to spend less money on marketing, and more on doing research to demonstrate that their products actually produce real benefits. . . . Companies offering a genuinely valuable service would prosper. Charlatans offering genetic gobbledygook would go to the wall' (quoted in Randerson, 2008).

A publicly accessible registry of information on the current state of genetic tests has several key advantages. First, while regulatory agencies attempt to sort out an appropriate risk-based approach for genetic tests, the immediate creation of a registry would put information in the hands of interested consumers, health care professionals and others. Second, putting more information in the public domain would compensate for the jurisdictional limits of regulatory authority. In other words, restrictions or prohibitions on genetic tests in a particular state or country often do not prevent consumers from purchasing a test from a company that operates in another jurisdiction. Finally, despite the attention to DTC genetic tests by government agencies, professional associations and non-government organizations, direct sale of genetic tests is arguably not an urgent consumer protection issue. Consumers are not rushing in large numbers to buy DTC genetic tests, especially nutrigenetic tests. In a US survey (Goddard et al., 2007) of over 5000 respondents, only 14% were aware of nutrigenetic tests and 0.6% had purchased a test. In Canada, the numbers are even lower (see Chapter 10). Yet, at the early stage of new technology adoption, clear playing rules benefit companies and consumers both. Such rules do not necessarily have to take the form of laws and regulations enacted by governments, but codes of fair advertising practices and genetic test information registries developed jointly among the DTC testing industry, government agencies, consumer organizations, health professional groups and other stakeholders may be the best approach in a field that is still in flux.

CONCLUSION

An increasingly wide variety of genetics tests are marketed and sold directly to consumers via the Internet. Defining and understanding the purpose, harms, benefits and implications of such tests are relevant to regulation, yet there is currently no consensus on how best to regulate this evolving field. Regulation commensurate with risk is desirable, but difficult to achieve where risks are not readily apparent or measurable. Discourse about DTC genetic tests should be careful not to make assumptions about how individuals will react to and use test results and presume that people need to be protected from information. Indeed, calls for blanket prohibitions or restrictions on genetic tests have attenuated in some quarters and have been replaced with acknowledgment that a single regulatory approach for all DTC genetic tests is not feasible or warranted.

Existing regulatory tools, such as truth-in-advertising rules, can be used to protect consumers from the worst players in the DTC genetic testing industry. But protecting consumers from unscrupulous firms is just one part of the regulatory task. Companies that desire a long-term, respected market presence in genetic testing have an interest in regulatory compliance. Governmental, professional and advocacy organizations concerned with DTC genetic testing should work to develop appropriate risk-based strategies for regulation and focus particularly on coordinated efforts to promote improved information disclosure about the strengths and limits of the current generation of DTC genetic tests.

References

American College of Medical Genetics (2004). ACMG Statement on Direct-to-Consumer Genetic Testing. *Genet Med* 6:60.

American College of Medical Genetics (2008). ACMG Statement on Direct-to-Consumer Genetic Testing. Online: www.acmg.net.

American Medical Association (2008). News Release: AMA adopts new policies at Annual Meeting. Online: http://www.ama-assn.org/ama/pub/category/18690.html.

(New policies adopted include recommendations on DTC advertising and provision of genetic testing.)

American Society of Human Genetics (2007). Statement on Direct-to-Consumer Genetic Testing in the United States. *Am J Hum Genet* 81:635.

Avey, L. (2008). Presentation to July 8, 2008 meeting of the Secretary's Advisory Committee on Genetics, Health and Society. Online: http://www4.od.nih.gov/oba/SACGHS/meetings/2008Jul/Avey.pdf.

Burke, W., Pinsky, L.E. and Press, N.A. (2001). Categorizing genetic tests to identify their ethical, legal and social implications. *Am J Med Genet (Semin Med Genet)* 106: 233–40.

Council of Europe (2008). Additional protocol to the convention on human rights and biomedicine, concerning genetic testing for health purposes. Online: http://conventions.coe.int/Treaty/EN/Treaties/Html/TestGen.htm.

Editorial (2007). Risky business. *Nature Genetics* 39: 1415.

Entwistle, P.A. (1976). Do-it-yourself pregnancy tests: The tip of the iceberg? *Am J Publ Hlth* 66:1108–9

Fabbri, W.O. (1995). Home HIV testing and conflicts with state HIV testing regulations. *Am J Law Med* 21:419–44.

Federal Trade Commission (2006). At-home genetic tests: a healthy dose of skepticism may be the best prescription. Online: http://www.ftc.gov/bcp/edu/pubs/consumer/health/hea02.shtm.

Federal Trade Commission (1999). Press release: 'Operation Cure.all' targets Internet health fraud (24 June 1999). Online: http://www.ftc.gov/opa/1999/06/opcureall.shtm.

Food and Drug Administration (2008). Donor screening assays for infectious agents and HIV diagnostic assays. Website updated March 18, 2008. Online: www.fda.gov/Cber/products/testkits.htm.

Furness, P., Zimmern, R., Wright, C. and Adams, M. (2008). The evaluation of diagnostic laboratory tests and complex biomarkers: summary of a diagnostic summit. Joint report of the Public Health Genomics Foundation and the Royal College of Pathologists. Online: www.phgfoundation.org/file/3998/.

Genetics and Public Policy Center (2007). News release: comparison of state laws for direct-to-consumer testing (6 July 2007), available online: Genetics and Public Policy Center http://www.dnapolicy.org/news.release.php?action=detail&pressrelease_id=81. See also Genetics and Public Policy Center (2007). Survey of direct-to-consumer testing statutes and regulations. Washington, D.C. Online: http://www.dnapolicy.org/resources/DTCStateLawChart.pdf.

Geransar, R. and Einsiedel, E. (2008). Evaluating online direct-to-consumer marketing of genetic tests: Informed choices or buyers beware? *Genet Test* 12(1):13–23

Goddard, K.A.B., et al. (2007). Awareness and use of direct-to-consumer nutrigenomic tests, United States, 2006. *Genet Med* 9:510–17.

Government Accountability Office (2006). Nutrigenetic testing: tests purchased from four web sites mislead consumers. Online: http://www.gao.gov/new.items/d06977t.pdf.

Heshka, J.T., et al. (2008). A systematic review of perceived risks, psychological and behavioral impacts of genetic testing. *Genet Med* 10:19–32.

Hogarth, S., Melzer, D. and Zimmern, R. (2005). The regulation of commercial genetic testing services in the UK: a briefing for the Human Genetics Commission. Online: http://www.phpc.cam.ac.uk/epg/dtc.pdf.

Human Genetics Commission (2003). Genes direct: ensuring the effective oversight of genetic tests supplied directly to the public. Online: http://www.dh.gov.uk/en/Publicationsandstatistics/Publications/PublicationsPolicyAndGuidance/DH_4084423.

Human Genetics Commission (2007). More genes direct: a report on new developments in the availability, marketing and regulation of genetic tests supplied directly to the public. Online: http://www.hgc.gov.uk/UploadDocs/DocPub/Document/More%20Genes%20Direct.pdf.

Janssens, A.C.J.W., et al. (2008). A critical appraisal of the scientific basis of commercial genomic profiles used to assess health risks and personalize health interventions. *Am J Hum Genet* 82:593–99.

Melzer, D., Hogarth, S., Liddell, K., Ling, T., Sanderson, S. and Zimmern, R. (2008). Evidence and evaluation: building public trust in genetic tests for common diseases. Research report of the Public Health Genomics Foundation and the Peninsula Medical School. Available online: www.phgfoundation.org/file/4003/.

Pearson, H. (2008). 'Breaking the news' in Genetic Testing for Everyone. *Nature* 453:570.

Randerson, J. (2008). Scientists urge more regulation of DIY kits for health checks. The Guardian 11 March 2008. Online: http://www.guardian.co.uk/society/2008/mar/11/health.medicalresearch.

Ries, N.M. (2008). Regulating nutrigenetic tests: an international comparative analysis. *Hlth Law Rev* 16:9–20.

Salbu, S.R. (1994). HIV home testing and the FDA: the case for regulatory restraint. 46 Hastings L.J. 403-457.

Schmidt, C. (2008). Regulators weigh risks of consumer genetic tests. *Nat Biotechnol* 26:145.

Secretary's Advisory Committee on Genetics, Health and Society (2008). US system of oversight of genetic testing: a response to the charge of the Secretary of Health and Human Services. Online: http://oba.od.nih.gov/oba/SACGHS/reports/SACGHS_oversight_report.pdf.

Secretary's Advisory Committee on Genetics, Health and Society (2007). US system of oversight of genetic testing: a response to the charge of the secretary of HHS draft report of the Secretary's Advisory Committee on Genetics, Health, and Society. Department of Health and Human Services. Draft report made available for public comment between November 5 and December 21, 2007.

Secretary's Advisory Committee on Genetic Testing (2000). Enhancing the oversight of genetic testing: recommendations of the SACGT. Online: http://www4.od.nih.gov/oba/sacgt/reports/oversight_report.pdf.

Secretary's Advisory Committee on Genetic Testing (2001). Development of a classification methodology for genetic tests: conclusions and recommendations of the secretary's advisory committee on genetic testing. Online: http://www4.od.nih.gov/oba/sacgt/reports/Addendum_final.pdf.

Wadman, M. (2008). Gene-testing firms face legal battle. *Nature* 453:1148–9.

Walensky, R.P. and Paltiel, A.D. (2006). Rapid HIV testing at home: does it solve a problem or create one? *Ann Intern Med* 145:459–62

CHAPTER

6

The Impact of Genomics on Innovation in Foods and Drugs: Can Canadian Law Step Up to the Challenge?

Karine Morin

OUTLINE

Nutrition and Genomics
ISBN: 978-0-12-374125-7

SUMMARY

Genomics is affecting innovation in food science and as well as pharmaceutical drug research and development. However, before new food products or drugs derived from these scientific advances will be able to enter the marketplace, they will be subjected to regulatory oversight. Using Canada as a case study, this chapter examines how nutrigenomics and pharmacogenomics may be driving changes in the food and pharmaceutical sectors and how personalized nutrition and health care in turn could transform marketing practices, bolstering the health benefits of foods and claims of efficacy for drugs. It also contrasts the regulatory regimes that govern these products, which share a common goal of protecting consumers by relying on scientific evidence to demonstrate safety and efficacy but not the experience of weighing risks and benefits among different segments of the population, as foods are generally available to all consumers without restrictions. These similarities and differences raise a critical question: will Canadian food and drug regulations accommodate genomic innovation or are changes necessary? The chapter ends with a discussion of the outlook for reforms that would encourage genomic innovation for the benefit of health conscious consumers and patients undergoing various drug regimens.

INTRODUCTION

Consumers may think of food and pharmaceutical drugs as very different products but, in Canada, they traditionally have been dealt under a common regulatory framework – the Food and Drugs Act. More stringent regulatory controls are imposed on drugs than on foods, but the development of new food products that have demonstrated impacts on health has begun to blur the distinction between food and drugs. Also, advances in genomics are beginning to impact food and nutritional sciences in ways similar to the impact on pharmacology and therapeutic innovations. To the extent nutrients and drugs affect genes in comparable ways, the food and pharmaceutical sectors may begin to resemble each other more as they adapt to the impact of genomics on the development and marketing of their respective products. This prospect raises the need to evaluate the current regulatory framework in Canada to determine whether it is adequate to oversee genomic innovation or whether it may act as a deterrent.

This chapter provides an overview of the processes involved in developing new foods and new drugs and the potential impact of genomics in this area. Using the Canadian legal and policy context as an illustrative case example, the chapter will review the regulatory framework that currently governs these products and, particularly, the provisions related to health claims linked to foods. Finally, it will examine whether the regulatory regime will be able to handle genomic advances in the development of foods and drugs or whether changes will be required.

FOOD, DRUGS AND GENOMIC PROMISES

The hope that biotechnology would impact health in new ways has been at the core of the Human Genome Project and persists to this day. Early expectations were that genomics could enhance drug development and help the pharmaceutical sector overcome some of the challenges in identifying new chemical entities as good candidates for therapeutic drugs. After watching the number of new drug applications submitted to it steadily decline over the past decade, the US Food and Drug Administration (FDA) issued a report in 2003 recognizing genomics as a means of circumventing some barriers to successful drug development (Lesko and Woodcock, 2004).

Multinational food companies are also paying attention to genomics, ranging from Nestle's interest in the genomics of fermentation (McNally, 2007) to the Dutch food giant DSM's venture investment in Sciona, a leader in the field of nutritional genomics (DSM, 2004). Shiseido, Japan's leader in cosmetics, extended its research portfolio in the late 1980s and started selling hypoallergenic rice in 1991, soon recognized as a 'Food for Specified Health Use' by that country's Ministry of Health and Welfare. A few years later, European researchers developed a strain of rice yielding provitamin A, leading the way for the production of other plant-based therapies (Simon and Kotler, 2003).

These trends underlie the expansion of pharmacogenomics and nutrigenomics and may help bring more products to the market that are the result of genomic innovation. Pharmacogenomics, broadly understood, is the study of genetic variations that affect drug response in terms of safety as well as efficacy. It can lead to clinical trials for new drugs to be conducted in relatively smaller cohorts of patients selected according to their genetic profile. In the clinical setting, it can enable patients to receive medicines from which they are most likely to derive benefits on the basis of their genotype or, conversely, can prevent the prescription of products likely to cause adverse effect. Moreover, it could affect the evaluation of the safety profile of medicines already on the market (Breckenridge et al., 2004). Pharmacogenomics, therefore, comes into play in the discovery of drugs, in the evaluation of safety and efficacy when drugs are in development and in the assessment of a new drug at the time of market approval (Webster et al., 2004).

The term 'pharmacogenetics', which often is used to refer to individual variation in the reaction to a drug being linked to genetically inherited traits, has roots in scientific knowledge dating back to the 1950s. The concept of 'pharmacogenomics' emerged in the late 1990s and is often associated with the application of genomics to drug discovery, but more specifically extends to the science and technologies associated with dividing populations or patients into groups on the basis of genotype and biological response to pharmaceutical drugs (Hopkins et al., 2006).

Nutrigenomics similarly refers to genetic variations, this time in relation to the consumption of nutrients. Historically, nutritional science treated everyone as genetically identical, despite observations that some individuals whose diets appear to predispose them to chronic disease remain healthy and others consuming similar diets develop the anticipated conditions. Nutrigenomics distinguishes itself in that responses to diet are analyzed according to genotype. It integrates nutrition, molecular biology, genomics and bioinformatics to identify and understand population and individual differences in gene expression in response to diet (Fogg-Johnson and Kaput, 2003).

It is important to note that the interactions between genotype and food consumption has to be conceptualized both in terms of the 'different effect of an environmental exposure on disease risk in persons with different genotypes' and the 'different effect of a genotype on disease in persons with different environmental exposures' (Kaput, 2007). This research is complicated by:

1. the diversity in genetic makeup of humans
2. the complexity of the chemicals that constitute naturally occurring foods and how they are prepared for consumption, and
3. the many ways that metabolism can be altered to produce disease. For example, some 300 genes have been linked to obesity and at least 150 gene variants have been implicated in type 2 diabetes (Horowitz, 2005).

Beyond the biological complexity, research in nutritional sciences is further undermined by some of the methodological limitations, including the shortcomings of relying on food frequency questionnaires (Tucker, 2007).

Similarities between pharmacogenomics and nutrigenomics are readily evident: they both aim to customize therapy and are likely to result in market segmentation based on personalized criteria (Fogg-Johnson and Kaput, 2003). However, from the standpoint of development, the pharmaceutical industry's approach of 'one drug, one target' cannot be replicated when it comes to foods. Diet is comprised of a multitude of nutrients, each involved in regulating disparate biological processes, which means that diet is a 'multiparametric approach' to preserving or optimizing health (Mutch et al., 2005).

The common promise of pharmacogenomics and nutrigenomics, upon which food and drug manufacturers intend to capitalize, is that scientific evidence will bolster claims made in relation to product efficacy. It is not surprising then that the public and regulatory sectors are not ignoring the trend. For example, the Public Health Agency of Canada recently recognized that:

> . . .the knowledge from advances in biotechnology and genome-based research can be applied to prevent disease and improve the health of populations. The Agency will therefore continue to form collaborative partnerships with national and international science and policy communities to translate rapidly evolving knowledge to improve health and reduce the impact of both chronic and infectious diseases (Public Health Agency of Canada, 2007).

More than partnerships will be needed to bring about changes that patients and consumers can experience directly: new products derived from genomic technology will only reach the market once they successfully meet regulatory requirements. Yet, food and drug law is rooted in a pre-genomic paradigm. Assessing how this will influence the impact of genomic innovation requires a review of current food and drug law, but also some understanding of current practices in food and drug research and development.

FOOD AND DRUG DEVELOPMENT

Food and drugs have significant similarities: both are 'consumed' and often ingested; both raise safety concerns, including worries about allergic or adverse reactions; both have the potential to affect health and well-being. A biomedical definition of food as material that contains essential nutrients (e.g. carbohydrates, fats, proteins, vitamins or minerals) and is ingested to produce energy, stimulate growth and maintain life, illustrates even further the convergence toward drugs (Chadwick, 2000). At the same time, food and drugs continue to be regarded as quite different: whereas food helps to maintain life and health, drugs are needed when one is afflicted by disease. The public generally perceives that foods are consumed for wellness whereas drugs are used to fight illness (Health Canada, 1998a) so we expect a drug will have some therapeutic effect (i.e. that it will work) but do not have that same expectation regarding foods. While both foods and drugs have been evaluated in terms of their safety and quality, only drugs were tested for efficacy – until recent innovations in foods (Chadwick, 2000).

A trend in claims of health benefits stemming from foods took flight with the 'Great Oat Rush' of the late 1980s, when companies quickly responded to reports linking oat bran consumption to cholesterol reduction and developed a multitude of products containing oat bran: 44 such products were created in 1988 and 218 in 1989 (Best, 1991). Kellogg's All-Bran® cereal was an early pioneer of the overall trend with advertisements in the mid-1980s that highlighted a National Cancer Institute recommendation that increased consumption of dietary fiber might reduce the risk of cancer (Burros, 1989). Overall, 2 million households were spurred to begin consuming high-fiber cereals, as industry responded by adding an average of 7% dietary fiber content to breakfast cereals (Best, 1991).

The Development of New Food Products

Research and development in the food industry is expensive and risky. In the late 1980s, the launch of a new food product cost approximately $54 million and that of a soft drink nearly $100 million. At the same time, it is estimated that only one product in ten ever makes it as far as consumer testing and only 10% of those ever makes it to market (Best, 1991). A decade ago, it was estimated that this failure rate represented losses of $12 billion annually for the food industry (Rudolph, 1995).

The difficulty in creating new markets for food is due, at least in part, to basic demographics: theoretically, food consumption can grow only as fast as the population (Best, 1991). In mature food markets of high-income countries like Canada, nutritional needs generally can be satisfied with available products and consumers do not necessarily increase expenditures on food commensurate with increases in their incomes (Veeman, 2002). Manufacturers' market shares and revenues grow principally from adding value to food products. Over many decades, through innovation in processing and packaging, manufacturers had focused on providing greater convenience, introducing ever more products for at-home consumption that are easy to prepare.

Foods that alleviate or prevent chronic diseases potentially represent the next trend in 'added-value', which could result in new developments and revenue growth (Veeman, 2002). This may depend in part on sociodemographic and economic changes associated with an aging and affluent population. It also would represent a new research and development orientation for food manufacturers who generally face low profit margins and spend on average less than 1% on R&D (Graf and Saguy, 1991). Compared to the pharmaceutical industry's 10% investment, this makes the food product industry one notorious for its low level of R&D spending. While pharmaceutical firms can rely on the value of a patent, which flows from claims of efficacy, the returns on any innovation in the food industry is contingent not only on a health claim but also convenience, taste and price (Massoud et al., 2001).

Consumers are the drivers of the food manufacturing industry innovation. New food product ideas originate from their perceived needs and are further conceptualized in terms of how to satisfy those perceptions (Jewson, 1991). The next steps in development rely directly on consumers who test and evaluate new foods. As test consumers voice what they want (for example: ketchup that is rich in tomato flavor, not too acidic, without water at the surface, which pours without scattering and is easy to squeeze), their feedback is translated into how a desired good can be brought to production (Dekker and Linnemann, 1998). Ultimately, a feasibility evaluation needs to be conducted, which is largely influenced by the formulation, i.e. the ingredients and recipe (Graf and Saguy, 1991). It is interesting to note that while past efforts had focused on alleviating diseases due to malnutrition through the fortification of foods by adding certain ingredients such as minerals and vitamins, today the emphasis is on alleviating the risk of chronic disease by removing ingredients such as fats, salt and sugars. Finally, manufacturers cannot neglect health safety; to this end, formulation, processing and packaging each must take into account food conservation and the avoidance of spoilage.

As nutritional science continues to make inroads in the prevention of disease and the promotion of health, biotechnology can play an increasing role in meeting American consumers' fixations on health and nutrition. Yet, food manufacturers face a significant challenge: despite the fact that more than 80% of adults 64 years and older have a chronic health condition, and despite growth in 'healthy' foods, there has also been a per capita increase in the consumption of salted snack foods and high calorie confections and desserts (Best, 1991). In other words, despite marketers' belief that consumers are on a wellness kick, there is a persistent increase in obesity rates. According to one analyst, manufacturers must be prepared to deal with consumers' fickleness. This means that when new products with health benefits are introduced, manufacturers face significant marketing costs because consumers have to be educated on the value of the innovation (Fuller, 2005).

Parallel to the manufacturers' efforts are those of government regulators. Since 1998, Health Canada has been committed to making chronic disease prevention a key objective of nutritional recommendation (Health Canada, 1998a). Accordingly, it recognized that consumers should be able to benefit from the full range of products that could contribute to health, including products such as functional foods and nutraceuticals. In turn, it became necessary to ensure that health claims, at that time undefined by Canadian law, would be 'supported by information that is clearly stated, substantiated, truthful, not misleading and not likely to lead to harm' (Health Canada, 1998a).

Until the introduction of this new policy, health claims in relation to foods had been prohibited. In crafting a regulatory exception, it could be expected that strict specifications would be established. In the end, Health Canada allowed claims that either related to 'structure/function' or referred to risk reduction, but did not permit therapeutic claims, concluding that such statements should be exclusive to drugs (Health Canada, 1998a).

With this policy in place and advances in nutritional genomics, innovative companies should be able to offer foods that enable consumers to live healthier and potentially longer lives simply by changing and monitoring their diets (Massoud et al., 2001). This would represent a significant shift:

> from today's world in which the dominant players leverage a low-margin, commodity business with ultra-efficient distribution systems and huge economies of scale to one with a growing number of smaller niches for value-added products and services meant for particular groups of consumers with similar health profiles (Massoud et al., 2001).

But the ultimate impact in the marketplace would depend on whether consumers learn about the connections between specific nutrients and their own genetic makeup and whether they would change their habits in light of this information (Cain and Schmid, 2003).

Drug Development

Medicinal drugs were once little more than herbs that underwent minimal processing, but they are now the product of the complex processing of biological and chemical substances, administered to prevent, alleviate or cure a disease, relieve a symptom or modify bodily functions in some way (Dukes, 2006). Because pharmaceutical innovation faces many challenges, including high R&D costs and long development times (Gassman et al., 2004), the industry strives to improve the early phase in drug discovery by identifying promising chemical entities as early as possible. This requires screening vast, randomized chemical libraries against a small number of pharmacologically relevant biological targets (Davidov et al., 2003), a process now aided by high throughput technology that allows for the automation of much of the discovery function, promoting a more comprehensive and consistent screening process. Yet, if genomics has the potential to accelerate the process of discovering novel targets, a lack of maturity still characterizes the technologies that play critical roles in determining the biological functions of targets and in translating knowledge into drugs (Gassman et al., 2004).

There is some uncertainty as to the economic impact most likely to result from these genomic advances. For instance, a Lehman Brothers study expected genomics to increase overall R&D costs and the average cost per new chemical entity. It found that 'despite the current need for better target validation through functional genomics, these technologies are unlikely to add value in the near term. These technologies are simply not yet robust enough to yield truly validated targets' (Charles River Associates,

2004). The Boston Consulting Group predicted that the implementation of genomics could increase efficiency or reduce failure rates in the future, such that companies on average could realize savings of nearly $300 million and 2 years per drug. However, a counter-balancing effect could be anticipated, as other forms of quality controls would need to be established to understand target function and develop appropriate assays in target validation and screening (Charles River Associates, 2004). Some researchers predict there will be a need for 'large-scale prospective studies that measure genetics and other biomarkers over time and follow up with patients for long-term outcomes. Producing knowledge this way will be costly and time-consuming in the near future but could enable smarter, faster therapy and diagnostic design in the long run' (Garrison and Finley Austin, 2006). In essence, if there is a discernable consensus that implementing new development technologies increases the costs of research in the short-term, it remains unclear whether these additional costs will be paid back and overall cost of development would be significantly lowered (Charles River Associates, 2004).

Beyond genomics-based R&D, other analysts foresee a transformation of marketing provoked by the new paradigm of pharmacogenomics as a 'fusion of diagnostics and therapeutics, with the joint development of tests and drugs based on the same molecular target' (Simon and Kotler, 2003). Marketing would be evidence-based, where a clear differential advantage is identified well before the launch of a product. But this convergence of diagnostics and therapeutics would also mean the integration of two separate business models that differ in terms of research investments, discovery timelines, regulatory approval processes and channels, as well as customers (Simon and Kotler, 2003). Moreover, in terms of pricing and reimbursements, diagnostics are usually characterized as 'resource- or cost-based, rather than value-based' (Garrison and Finley Austin, 2006).

FOOD AND DRUG LAW

To evaluate how the law will respond to products that are the result of genomic innovation in the food and drug sectors, it is important to be familiar with the current regulatory regime. Originally, food laws were narrowly concerned with aspects of fraud and deceit, and drug laws with potential harmful effects. In Canada, these matters fell under the federal government's jurisdiction over criminal law matters (Curran, 1953). Consumer protection laws have evolved into risk-management approaches that balance benefits and risks in terms of consumers' access to products and industry's economic well-being. The overarching goal is to ensure safety and enhance health without undue barriers to competition (Health Canada, 1998a).

Food and drug law focuses primarily on the safety and quality of products and, to the extent that specific claims are made about their use, efficacy. As mentioned at the outset of this chapter, foods and drugs are regulated under a common statutory framework in Canada and a product's classification as a food or a drug determines the specific provisions that apply. New products, however, create classification difficulties: beyond traditional foods, there are now organic foods, novel foods, functional foods, genetically modified foods, as well as nutraceuticals, medicinal herbs, supplements, vitamins and minerals, all consumed as part of ordinary diets, often unrelated to any diagnosed ailment, but with expectations of health benefits (Veeman, 2002).

Food Law

Foods were long sold in bulk, creating risk of adulteration with the mixing of inferior or cheaper substances, reducing the quality, or even causing the food to be injurious (Curran, 1953). In 1934, Canada's Food and Drugs Act underwent an important change by including a provision that prohibited the sale of a remedy marketed as a treatment for any health condition listed in the statute (e.g. alcoholism, cancer, diabetes, epilepsy, heart disease, obesity and sexual impotence). It is important to note that the restriction referred neither to a food, defined as including every article used for food or drink or any ingredient intended for mixing with the food or drink, nor a drug, merely defined as medicines for internal or external use. Since then, the relevant provision has been modified to state that no person shall advertise or sell any food or drug as a treatment, preventative or cure for any of the diseases, disorders or abnormal physical states referred to in a schedule of the act.

Other key provisions stipulate that food shall not be sold if it is poisonous or includes a harmful substance; is unfit for human consumption; consists of any filthy, decomposed or diseased animal or vegetable substance; is adulterated; or prepared under unsanitary conditions. Nor shall food be sold or advertised in a manner that is false, misleading or deceptive or is likely to create an erroneous impression regarding the food's character, value, quantity, composition, merit or safety. Essentially, the emphasis has been in regulating safety and information about food, rather than its function or efficacy (Cain and Schmid, 2003). It has been remarked that food manufacturers are held to high safety standards because food choice is discretionary. In contrast, drugs may present potential adverse effects, but these are nonetheless tolerated because therapeutic benefits outweigh the risks for those who are unwell (Health Canada, 2000a).

Interestingly, food safety does not necessarily entail pre-market evaluation and the vast majority of foods available to Canadian consumers are not scrutinized by regulators before they are available for purchase.

Only products that fall in specific categories (infant formulas, food additives, irradiated foods) or those designated as novel foods undergo review. More commonly, to ensure safety, manufacturers are limited to the use of ingredients according to standards set out in the regulations, which specify composition, strength, potency, purity, quality or other property specific to the relevant food (Health Canada, 1998a). The regulations establish limits on the presence of toxic chemical residues, microbiological hazards or extraneous matters and standardize the composition of foods (Fuller, 2005). Importantly, tests that help evaluate toxicity are completely distinct from tests that would evaluate the health benefits that may accrue by consuming certain foods. While the Canadian Food Inspection Agency is responsible for handling food safety, the responsibility of promoting healthy eating falls under a separate branch of the federal government, the Office of Nutrition Policy and Promotion.

This dichotomy between safety and health benefits constitutes a challenge for manufacturers intent on developing value-added products to enhance consumer health. Currently, such niche markets may be underserved due to a lack of comprehensive studies on consumer preferences and motivations for purchasing foods (Veeman, 2002) but it is also due to long-standing policy and regulatory hurdles.

Health Claims in Canadian Law

Food shoppers are driven by taste and convenience and often lack the knowledge that would inform nutritious choices. This situation is one ripe for misrepresentation, which in part justifies regulatory controls to mitigate market failures attributable to asymmetric information. However, manufacturers may soon want to provide more information for consumers who are better educated about nutrition or who, through nutrigenomic testing, know more about their nutritional needs. Policy on health-oriented food products is relatively recent and will be challenged to keep up with related scientific developments (Veeman, 2002).

Historically, food labels have listed ingredients; also, somewhat particular to Canadian food legislation, they have included 'grade standards', which were indicative of a certain level of quality and were usually reflected in price differences. Yet, a higher grade was not necessarily synonymous with higher nutritional value (Curran, 1953). The most recent food labeling regulatory reform has brought about significant changes. With the goal of giving consumers more information that could help them make healthier choices, Health Canada has determined that health claims should be allowed but only if they are supported by substantial evidence. A health claim is considered valid if its efficacy (can the product produce the health effect it claims) as well its effectiveness (does the product produce the effect it claims) are demonstrated. This is a matter of establishing a link between the desired effect and the consumption of the product or relevant bioactive (Health Canada, 2000a).

Holding public health and consumer confidence as paramount, the framework developed to evaluate the evidence behind health claims was also intended to make it possible for industry to develop products at reasonable costs without more trade restrictions than necessary and also with a level-playing field across product categories (Health Canada, 2000a). To this end, Health Canada developed two alternative health claim models: one known as the 'generic authorization', the other a 'product-specific authorization'. The generic authorization model, first issued in 2000, permits five risk reduction claims that are diet-related: sodium and hypertension, calcium and osteoporosis, saturated and trans fat and cholesterol and coronary heart disease, fruits and vegetables and cancer, and sugar alcohols and dental caries (Health Canada, 2000b). This framework was included along with other changes to food labeling as part of amendments made to the Food and Drug Regulations in 2003.

A product-specific authorization model was issued in October 2001 to cover foods intended either to affect a function or body structure beyond maintenance of good health or normal growth and development or to reduce the risk of or facilitate the dietary management of disease (Health Canada, 2001). To aid food manufacturers wishing to market products touting health benefits, Health Canada released additional guidance the following year (Health Canada, 2002). Claims can only be made after a submission for a Claim Identification Number has been prepared in compliance with detailed standards of evidence.

Foods falling under the generic health claim model require less exhaustive documentation for review than foods falling under the product-specific claim model. Most notably, the latter requires the submission of product-specific human experimental studies to validate claims. It is worth remembering that the evidentiary documentation required by these regulations is not specifically related to the food products, but rather to the claims about their benefits. This is a situation similar to the one that exists in the USA, which led to the following remark:

> It is fairly easy to get permission from the FDA to distribute new food products to the public, yet strict regulations limit the health claims companies can make for their products. . . . However, once a new food product or ingredient is approved for making a particular claim, any other product that meets the same criteria can make the same claim. Consequently, food companies find difficulty justifying the enormous research costs necessary to prove to the FDA the validity of health claims (Massoud et al., 2001).

Health Canada has similarly acknowledged that product-specific claims do not imply health claim exclusivity; i.e. nothing prevents another manufacturer from making the same claim (Health Canada, 2000a), whether its product has already been on the market thus far without any claim, or whether a product is subsequently introduced with the same claim.

According to researchers assessing the economic rationale for the regulation of health claims, the evidentiary hurdle may be justifiable to remedy market failures arising from imperfect information on foods and the risk of misrepresentation, but the flip side is that prohibited claims represent missed opportunities for the public.

> Indeed there is a substantial body of evidence supporting the notion that provision of information on health effects through product labels not only brings about positive changes in consumer dietary choices, but also intensifies market competition among manufacturers for the supply and disclosure of valued product attributes, in turn enhancing consumer choices (Herath et al., 2006).

Pharmaceutical Drugs and Medical Devices

Similar to the concern regarding the adulteration of foods, early drug laws dealt with the possibility that their composition would be manipulated such that they would not conform to standards regarding potency, quality or purity set by the pharmacopeia of the time (Curran, 1953). With the development of more sophisticated methods to evaluate drugs and their effect in the late 19th and early 20th centuries, concerns related to efficacy were added to those of safety and truthful representation (Dukes, 2006). Today, Canada's Therapeutic Products Directorate assesses the benefits and risks of a drug prior to its approval for sale on the basis of the scientific evidence that a manufacturer presents regarding a product's safety, efficacy and quality (Health Canada, 2006).

Following in the footsteps of the US (Food and Drug Administration, 2005), Canada issued guidance that requires drug sponsors, as part of a clinical trial application, to submit available pharmacogenomics data that pertain to the characteristics of the investigational product or support its safety and/or efficacy, as well as data to support the design of the proposed clinical trial or to justify the proposed indication(s)/labeling (Health Canada, 2007). The guidance distinguishes between pharmacogenomic testing using either a test already licensed for sale in Canada or already authorized for investigational testing, or using a test that is not presently licensed or authorized. In the latter case, the sponsor must obtain an authorization for investigational testing. This entails submission of all available data that support the analytical validity of the test to be reviewed by the Medical Devices Bureau (Health Canada, 1998b). Accordingly, in addition to meeting the requirements regarding drug development, pharmaceutical and biotech companies must also be concerned with the regulations regarding diagnostic devices. This regulatory regime is quite different from the one related to drugs. Specifically, in Canada, as in the USA, the federal regulations only extend to devices that are sold. Because of this loophole, many diagnostic devices, including many genetic tests relevant to pharmacogenomics, are governed by laboratory accreditation provisions. Diagnostic devices that do fall within the scope of the medical devices regulations are evaluated according to a risk-based classification system under which pharmacogenomic tests are classified as Class III and require a pre-market scientific assessment of their safety and effectiveness (Health Canada, 1998c).

Overall, there is an uncoordinated regulatory landscape for drugs and diagnostics that will affect the advancement of pharmacogenomics. Similarly, there are many regulatory layers that apply to food manufacturers in terms of safety on one hand and in terms of the validity of health claims on the other, which will impact their ability to make claims based on nutrigenomic data. These challenges are clear illustrations of the byzantine regulatory world that awaits genomic innovations.

DOES GENOMICS REQUIRE REGULATORY INNOVATION?

Statutes and regulations are instruments to achieve policy goals (Rubin, 1991). They have long been viewed from the perspective of a battle opposing those who favor strong regulation to those who favor deregulation. The policy horizon may be broadened to include 'private regulation', in which conduct is prescribed by industry associations, firms, peers or individual consciences (Ayres and Braithwaite, 1992).

Within this new frame, calls for reform have advocated the need for regulations to be responsive in form to an industry's structure and in degree to its actors' behavior (Ayres and Braithwaite, 1992). New regulatory regimes should be designed to achieve desired outcomes without specific or rigid rules to be followed by all in the exact same manner. Instead, rules are intended to promote behaviors that seek to prevent problems before they occur. For example, incentives are offered for voluntary implementation of compliance systems. This reliance on persuasion and cooperation rather than deterrence is viewed as a new means of enticing firms toward policy goals. It is also intended to help rein back regulatory inflation (Parker, 2000) and avoid the proliferation of unnecessary, contradictory, or overly complex regulations.

Overstating the need for regulatory innovation is perhaps reminiscent of the concern raised by 'genohype' (Holtzman, 1999) or tendency to exaggerate claims attached to the human genome project and its aftermath (Fleising, 2001). In fact, some caution that the need for regulatory innovation should be invoked only when new (or newly constructed) problems arise or old ones persist (Black et al., 2005). In this light, do genomic advances regarding either the development of food and elaboration of health claims or the development of drugs and potential genetic-based labels present new regulatory problems or is this just a matter of regulatory hype?

Earlier in this chapter, genomics advances were described as presenting additional R&D challenges to both the food and pharmaceutical sectors, which may translate into the regulatory realm. For pharmaceutical companies, pharmacogenomics potentially holds the promise of reducing the cost of developing new drugs. Once pharmacogenomic interventions reach the market, however, the benefits to patients, namely reducing the risk of adverse effects or avoiding ineffective treatments, stem from products being targeted to limited segments of the population. This is in sharp contrast to current market opportunities where products can be offered to all patients with a given condition with the hope that they will experience a benefit. In 2006, the US National Research Council commented on the expected impacts of genomics:

> ... drug companies will come under increasing pressure to tailor therapies to individual groups of patients sharing a particular genomic/proteomic signature or fingerprint (as well as certain nongenetic traits). This sea change will first become apparent in the design and execution of clinical trials, in which genetic predispositions to therapeutic benefits and risks will be analyzed. The benefit to pharmaceutical manufacturers will be cheaper, faster clinical trials promising a higher likelihood of detecting a positive signal and a reduction in the number of adverse events, leading to more rapid approvals. Such advances will, however, come with a price for pharmaceutical and biotechnology companies. The advent of personalized medicine may well bring an end to the era of so-called blockbuster drugs, because product development will be restricted to smaller target patient populations. To say the least, the current economics of drug discovery, development, and marketing will change considerably (Merrill and Mazza, 2006).

Others (Gassman et al., 2004) echo the demise of the blockbuster drug but the inevitability of personalized medicine still raises doubts. If health care payers' concerns over drug safety, effectiveness and costs are likely to drive change, it is worth noting that 'the few cost-effectiveness analyses of [pharmacogenomic] interventions that have been conducted show inconclusive results as to whether such interventions are a relatively good value for society' (Phillips, 2006). Moreover, 'the move toward greater personalization of medicine might directly conflict with two other major concerns in the United States: growing spending on health care and pharmaceuticals, and access to care for underserved populations' (Phillips, 2006).

Although the Canadian health care system is primarily publicly financed, similar concerns regarding universal access to costly technologies do exist. In the context of pharmacogenomics, part of the solution may be to prioritize research that would yield substantial public health improvements by targeting complex diseases with strong genetic contribution, but where the ability to modify exposure or eliminate risk factors has been limited (Merikangas and Risch, 2003). Evidently, this would not require changes to current drug regulations but rather new priorities in research funding. So, apart from an economic assessment of the impact of pharmacogenomics, regulators focusing on safety, through the reduction of adverse events, and efficacy, through the identification of effective therapeutic drug interventions, may be confronted with a choice between mandating genomic data as new drug products are submitted for market approval, enticing pharmaceutical companies to submit such data, or else penalizing them for the failure to do so, for example through different degrees of liability protection.

With regard to the first option, some commentators already note that a much more nuanced approach would be necessary, as a drug that is highly efficacious across most of the population, shows little inter-individual variability in kinetics and dynamics and has a wide therapeutic index (or differential between a therapeutic dose and a toxic one) should probably not require pharmacogenomic data (Webster et al., 2004). As to providing liability protection to manufacturers who submit pharmacogenomic data, it could help prevent cases like the one in the USA, where patients with a particular genotype developed an incurable form of arthritis after being vaccinated for Lyme disease. They sued the maker of the vaccine, claiming that the product should have been labeled to indicate that those with such a genotype could suffer adverse events (Warner, 2000; Binzak, 2003).

To entice the development and marketing of pharmacogenetic interventions, regulators perhaps ought better to coordinate the evaluation of the data through the respective drug or medical device regulatory pathways. At last, this may require a more explicit recognition that diagnostic tests, including genetic tests developed within a laboratory and not sold as kits, must be regulated under the medical device regulations.

Turning to foods, the challenge is considerably different. Food manufacturers, as discussed above, have few incentives to engage in expensive R&D to substantiate health claims. To the extent that nutrigenomics provides data that could be applied to existing food products, however, there may not be expensive research to undertake; instead, manufacturers may simply be able to personalize health claims. Yet, to be attractive to consumers, such products would have to communicate the benefits in an effective manner (Herath et al., 2006). This is no small matter: relying on focus groups conducted for Health Canada to see how consumers would respond to the wording of claims, researchers found a general lack of knowledge in basic nutrition that hindered the comprehension of claims and wariness toward indefinite concepts used to describe health benefits (Jones and Bourque, 2003). To the extent that more education in nutrition would be necessary to complement the establishment of health claims, nutrigenomics may not address consumers' desires for 'guarantees' when it comes to claims but could expose the multifactorial nature of disease (Jones and Bourque, 2003).

To reward the efforts of manufacturers who present nutrigenomic data as part of product-specific claims, it has been suggested that they should be entitled to some shorter form of intellectual property protections than exist for drugs, as long as prices were not affected to the point of compromising access (Jones and Bourque, 2003). But current food regulations may not be ready for such changes. With the emphasis on safety, it is important to recognize that risks are conceived primarily in terms of harms that could arise immediately upon consumption, such as those that can result from ingesting contaminated foods. This is contrary to evolving knowledge regarding nutrition which, in part through nutrigenomics, makes clear that foods have a long-term impact on health; in fact, some foods can have a negative health impact even though each of their ingredients has been determined to be safe. Therefore, the long-term benefits of nutrition based on optimization as determined by nutrigenomic testing does not appear to fit into the current regulatory framework in terms of safety and risks.

From the perspective of health claims, all the attention continues to be placed on the risk of misleading consumers with claims that are not sufficiently proven, while there is a complete failure to recognize the risk of leaving consumers unaware of potential health benefits that exist in foods already available. In the absence of this information, consumers continue to make uninformed decisions, which often translate into unhealthy decisions. The persistence of this skewed vision is problematic (Health Canada, 2000a).

This outlook is illustrated in a report of the US Government Accountability Office (GAO) which recognized that, as a result of the introduction of the Dietary Supplement and Health Education Act (DSHEA) in 1994, a market for products with health claims had been able to flourish. Nevertheless, the agency persisted in questioning whether the law sufficiently ensured safety and prevented false or misleading claims (Veeman, 2002). The concerns over the validity of claims – or the strength of the evidence – also lead to the suggestion that, in order to maintain effective behavioral changes prompted by health claims, there must be absolute consumer confidence in the regulatory process through which health claims are generated. Moreover, because those changes are difficult to bring about, it has been suggested that emphasis should be placed on claims having major health impacts. Claims that have little or no effect should not be allowed, as they may dilute consumers' efforts to modify their behavior and undermine confidence in the regulatory process (Jones and Bourque, 2003).

Yet the potential in developing foods with health claims that are based on nutrigenomics could result in helping segments of the population take an active role in preventing chronic diseases or mitigating some of their impact. This could mean that regulators need to look differently at the generic and product-specific authorization models to determine whether health claims are allowable on the basis of nutrigenomic data. It is quite possible that few generic claims will succeed, but perhaps more likely that tailored claims could be permissible for specific products. Today, Health Canada contends that the complexity of foods is such that evidence required for foods with health claims can be higher than that for a new drug (Health Canada, 2000a). Once genomic data are commonly provided to regulators with respect to both gene–drug interactions and gene–food interactions, however, it would appear reasonable to expect that claims of efficacy on the basis of genomic data will be acknowledged and communicated to consumers.

References

Ayres, I. and Braithwaite, J. (1992). *Responsive regulation: transcending the deregulation debate.* Oxford University Press, New York.

Best, D. (1991). Designing new products from a market perspective. In *Food product development* (E. Graf and I.S. Saguy, eds.). Chapman & Hall, New York.

Binzak, B.A. (2003). How pharmacogenomics will impact the federal regulation of clinical trials and the new drug approval process. *Food Drug Law J* 58:103–27.

Black, J., Lodge, M. and Thatcher, M. (eds.) (2005). *Regulatory innovation: a comparative analysis.* Edward Edgar, Cheltenham.

Breckenridge, A., Lindpaintner, K., Lipton, P., Mcleod, H., Rothstein, M. and Wallace, H. (2004). Pharmacogenetics: ethical problems and solutions. *Nat Rev Genet* 5:676–80.

Burros, M. (1989). Flirting with health claim, Kellogg announces a cereal. *New York Times* August 31.

Cain, M. and Schmid, G. (2003). From nutrigenomic science to personalized nutrition: the market in 2010. Institute for the Future.

Chadwick, R. (2000). X-novel, natural, nutritious: towards a philosophy of food. *Proc Aristotelian Soc* 100:193–208.

Charles River Associates (2004). Innovation in the pharmaceutical sector: a study undertaken for the European Commission. (Online) Available at: http://www.crai.com/uploadedFiles/RELATING_MATERIALS/Publications/Consultant_publications/files/pub_4510.pdf.

Curran, R.E. (1953). *Canada's food and drug laws.* Commerce Clearing House, Inc, Chicago.

Davidov, E., Holland, J., Marple, E. and Naylor, S. (2003). Advancing drug discovery through systems biology. *Drug Discov Today* 8:175–83.

Dekker, M. and Linnemann, A.R. (1998). Product development in the food industry. In *Innovation of food production systems: product quality and consumer acceptance* (W. Jongen and M. Meulenberg, eds.). Wageningen Pers, Wageningen.

DSM (2004). DSM Venturing invests in Sciona Inc. (Online) Available at: http://www.dsm.com/en_US/html/media/press_releases/27_04_sciona.htm.

Dukes, G. (2006). *The law and ethics of the pharmaceutical industry.* Elsevier, Amsterdam.

Fleising, U. (2001). In search of genohype: a content analysis of biotechnology company documents. *New Genet Soc* 20:239–54.

Fogg-Johnson, N. and Kaput, J. (2003). Nutrigenomics: an emerging scientific discipline. *Food Technol* 57:60–7.

Food and Drug Administration (2005). Guidance for industry: pharmacogenomic data submissions. Rockville. (Online) Available at: http://www.fda.gov/Cber/gdlns/pharmdtasub.htm.

Fuller, G. (2005). *New food product development: from concept to marketplace.* CRC Press, Boca Raton.

Garrison, L.P. Jr. and Finley Austin, M.J. (2006). Linking pharmacogenomics-based diagnostics and drugs for personalized medicine. *Hlth Affairs* 25:1281–90.

Gassman, O., Reepmeyer, G. and von Zedtwitz, M. (2004). *Leading pharmaceutical innovation: trends and drivers for growth in pharmaceutical industry*. Springer-Verlag, Berlin.

Graf, E. and Saguy I.S. (eds.) (1991). R&D process. In *Food product development*. Chapman & Hall, New York.

Health Canada (1998a). Nutraceuticals/functional foods and health claims on food. Therapeutic Products Programme and the Food Directorate, Health Protection Branch, Health Canada, Ottawa. (Online) Available at: http://www.hc-sc.gc.ca/fn-an/alt_formats/hpfb-dgpsa/pdf/label-etiquet/nutra-funct_foods-nutra-fonct_aliment_e.pdf.

Health Canada (1998b). Preparation of an application for investigational testing – in vitro diagnostic devices (IVDD) – guidance document. Bureau Medical Devices, Therapeutics Products Directorate, Health Canada, Ottawa. (Online) Available at: http://www.hc-sc.gc.ca/dhp-mps/md-im/applic-demande/guide-ld/labl_etiq_ivd_div_e.html.

Health Canada (1998c). Preparation of a premarket review document for class iii and class iv device licence applications. Bureau Medical Devices Therapeutics Products Directorate, Health Canada, Ottawa. (Online) Available at: http://www.hc-sc.gc.ca/dhp-mps/md-im/applic-demande/guide-ld/prmkt2_precomm2_e.html.

Health Canada (2000a). Consultation document: standards of evidence for evaluating foods with health claims: a proposed framework. Bureau of Nutritional Sciences, Food Directorate, Health Protection Branch, Health Canada, Ottawa. (Online) Available at: http://www.hc-sc.gc.ca/fn-an/alt_formats/hpfb-dgpsa/pdf/label-etiquet/consultation_doc_e.pdf.

Health Canada (2000b). Consultation document on generic health claims. Health Protection Branch, Health Canada, Ottawa. Bureau of Nutritional Sciences, Food Directorate, Health Protection Branch, Health Canada, Ottawa. (Online) Available at: http://www.hc-sc.gc.ca/fn-an/alt_formats/hpfb-dgpsa/pdf/label-etiquet/health_claims-allegations_sante_e.pdf.

Health Canada (2001). Product-specific authorization of health claims for foods: a proposed regulatory framework. Bureau of Nutritional Sciences, Food Directorate, Health Protection Branch, Health Canada, Ottawa. (Online) Available at: http://www.hc-sc.gc.ca/fn-an/alt_formats/hpfb-dgpsa/pdf/label-etiquet/final_proposal-proposition_final_e.pdf.

Health Canada (2002). Interim guidance document: preparing a submission for foods with health claims incorporating standards of evidence for evaluating foods with health claims. Bureau of Nutritional Sciences, Food Directorate, Health Protection Branch, Health Canada, Ottawa. (Online) Available at: http://www.hc-sc.gc.ca/fn-an/alt_formats/hpfb-dgpsa/pdf/label-etiquet/abstract_guidance-orientation_resume_e.pdf.

Health Canada (2006). Access to therapeutic products: The regulatory process in Canada. Health Products and Food Branch, Health Canada, Ottawa. (Online) Available at: http://www.hc-sc.gc.ca/ahc-asc/alt_formats/hpfb-dgpsa/pdf/pubs/access-therapeutic_acces-therapeutique_e.pdf.

Health Canada (2007). Guidance document: submission of pharmacogenomic information. Health Products and Food Branch, Health Canada, Ottawa. (Online) Available at: http://www.hc-sc.gc.ca/dhp-mps/alt_formats/hpfb-dgpsa/pdf/brgtherap/pharmaco_guid_ld_2007-02_e.pdf.

Herath, D., Henson, S. and Cranfield, J. (2006). A note on the economic rationale for regulating health claims on functional foods and nutraceuticals: the case of Canada. *Hlth Law Rev* 15:23–32.

Holtzman, N.A. (1999). Are genetic tests adequately regulated? *Science* 286:409

Hopkins, M.M., Ibarreta, D., Gaisser, S., et al. (2006). Putting pharmacogenetics into practice. *Nat Biotechnol* 24:403–10.

Horowitz, S. (2005). Nutrigenomics and nutrigenetics: personalizing nutrition based on genes. *Alternat Compliment Thera* :115–19.

Jewson, D. (1991). Consumer research. In *Food product development* (E. Graf and I.S. Saguy, eds.). Chapman & Hall, New York.

Jones, P.J.H. and Bourque, C. (2003). Health claims on foods in Canada: toward successful implementation. *Can J Pub Hlth* 94:260–4.

Kaput, J. (2007). Nutrigenomics – 2006 update. *Clin Chem Lab Med* 45:279–87.

Lesko, L.J. and Woodcock, J. (2004). Translation of pharmacogenomics and pharmacogenetics: a regulatory perspective. *Nat Rev Drugs Discov* 3:763–9.

Massoud, M., Ragozin, H., Schmid, G. and Spalding, L. (2001). The future of nutrition: consumers engage with science. Institute for the Future.

McNally, A. (2007). Nestle teams up with genomics super-league. Nutraingredients.com – Europe. (Online) Available at: http://www.nutraingredients.com/news/ng.asp?n=81587-nestl-genomics-industrial-fermentation.

Merikangas, K.R. and Risch, N. (2003). Genomic priorities and public health. *Science* 302:599–601.

Merrill, S.A. and Mazza, A.-M. (eds.) (2006). *Reaping the benefits of genomic and proteomic research: intellectual property rights, innovation, and public health.* National Academies Press, Washington, DC.

Mutch, D.M., Wahli, W. and Williamson, G. (2005). Nutrigenomics and nutrigenetics: the emerging faces of nutrition. *FASEB J* 19:1602–16.

Parker, C. (2000). Reinventing regulation within the corporation: compliance-oriented regulatory innovation. *Administrat Soc* 32:529–65.

Phillips, K.A. (2006). The intersection of biotechnology and pharmacogenomics: health policy implications. *Hlth Affairs* 25:1271–80.

Public Health Agency of Canada (2007). Strategic plan: 2007–2012, information, knowledge, action. (Online) Available at: http://www.phac-aspc.gc.ca/publicat/2007/sp-ps/pdfs/PHAC_StratPlan_E_WEB.pdf.

Rubin, E.L. (1991). The concept of law and the new public law scholarship. *Michigan Law Rev* 89:792–836.

Rudolph, M.J. (1995). The food product development process. *Br Food J* 97:3–11.

Simon, F. and Kotler, P. (2003). *Building global biobrands: taking biotechnology to market.* Free Press, New York.

Tucker, K.L. (2007). Assessment of usual dietary intake in population studies of gene-diet interaction. *Nutr Metab Cardiovasc Dis* 17:74–81.

Veeman, M. (2002). Policy development for novel foods: issues and challenges for functional food. *Can J Agricult Econ/Rev Can Agroecon* 50:527–39.

Warner, S. (2000). 4 sue SmithKline, claiming Lyme vaccine making them sick. Philadelphia Inquirer June 12.

Webster, A., Martin, P., Lewis, G. and Smart, A. (2004). Integrating pharmacogenetics into society: in search of a model. *Nat Rev Genet* 5:663–9.

CHAPTER

7

Placing Healthy Eating in the Everyday Context: Towards an Action Approach of Gene-Based Personalized Nutrition Advice

Laura I. Bouwman and Cees van Woerkum

OUTLINE

Nutrition and Genomics
ISBN: 978-0-12-374125-7

SUMMARY

The incidence of diet-related diseases, likely associated with energy-dense and nutrient-poor diets, is increasing rapidly. Nutritional intervention aims to inform and motivate people to make healthier food choices and, as a result, help to prevent diseases associated with poor diets. In this chapter, we introduce the innovative approach of including knowledge about individual predisposition to diet-related illnesses – i.e. nutrigenomics information – to personalized nutrition interventions. While this new information may draw attention to nutritional genomics, it is not yet clear how this evidence will fit into the process of motivating healthful eating, particularly by changing behavior. This new nutritional evidence may give an impulse to such interventions. However, it is not known whether it will actually motivate healthier eating practices. This chapter elaborates on a new approach that may answer why current personalized interventions are not always successful and deliver an alternative way of designing these interventions, with specific emphasis on the integration of genetic knowledge.

INTRODUCTION

Incidence of diet-related diseases, likely associated with energy-dense and nutrient-poor diets, is increasing rapidly (WHO, 2004; Kreijl et al., 2006). Nutrition advice aims to inform and motivate healthier eating behavior. In this chapter, we introduce an approach to dietary counseling that incorporates nutrigenomic information. Our focus is on discussing the use of individual, genetic information about susceptibilities to diet-related diseases to develop personalized nutrition advice.

Nutrigenomics is an innovative field that studies the interaction between food, genes and health at the molecular level. A genetic test for vulnerability to diet-related illnesses such as cardiovascular disease could be added to a personal risk assessment, one that is currently comprised of indicators such as body mass index and blood cholesterol. Results of such tests could be used to increase individual awareness about healthy eating and to develop individually tailored dietary advice (DeBusk and Joffe, 2006). Yet nutrigenomics raises questions, mainly regarding how this advice can be embedded in a broader approach in which not only the nutritional evidence is personalized, but so, too, is the way people learn to adjust their daily life behavior in light of the advice. This is the starting point for this chapter: aiming at an integrated strategy that takes into account new biomedical innovation as well as recent insights about how people change their behavior.

Motivating change through a personalized approach

Personalized nutritional interventions differ from other health promotion approaches in two ways: first, the messages or strategies are intended for one particular person rather than for a group of people; and, second, those messages or strategies are based on individual assessments. The provision of personalized nutrition advice is no longer the sole domain of dieticians. The rapid developments in interactive computer technology (ICT) applications, particularly the Internet, allow for tailored interventions with large reach at relatively low cost (cf. Eng, 2004; Brug et al., 2005). The interventions use computer programs to collect data about an individual's dietary intake, health indicators such as body mass index and psychosocial factors. Users receive personalized feedback about their current risk of developing diet-related illnesses and advice about how to reduce this risk by modifying their eating practices to accord with healthy eating guidelines.

Studies have shown that such personalized advice is more effective than generic messages in motivating individuals to adopt healthier eating behavior. Personalized interventions have been used to induce changes in smoking, diet and physical activity (Kreuter and Stretcher, 1996; Brug et al., 1998; Curry et al., 2005). In a systematic review of studies on computer-tailored nutrition and physical activity advice, Kroeze et al. (2006) found strong evidence for the effectiveness of computer-based, personalized interventions, especially in motivating reductions in dietary fat intake. Another review was less enthusiastic, concluding that current evidence is insufficient to conclude that computer-tailored interventions are superior to other interventions (DeNooijer et al., 2005).

The innovations of nutrigenomics and computer-tailored dietary advice within the context of behavior change theories will be evaluated. Based on this evaluation, this chapter will elaborate on a new approach towards motivating healthy eating. This approach may provide answers to questions about why current personalized interventions are not always successful and it may support the development of alternative ways of designing these interventions.

THE THEORETICAL BASIS FOR INDIVIDUAL BEHAVIOR CHANGE

Nutrition interventions are most likely to succeed if they are based on a clear understanding of eating behavior. Theories of health behavior are important to explain and understand healthy eating objectives and to indicate ways to achieve behavior change. Theories that aim to explain and predict individual eating behavior identify intrapersonal factors such as knowledge, attitudes, beliefs, motivation, self-efficacy and skills. All these factors are subject to change. For instance, the health belief model (Janz and Becker, 1984), which concerns individual perceptions about risks of unhealthy eating and the effectiveness of healthy eating advice, is frequently used to develop messages to persuade individuals to adopt healthier eating practices. Other valuable theories address the processes by which people take in and use information in their decision-making, such as Weinstein's Precaution Adoption Model (Weinstein, 1988). This model combines concepts from adoption processes of new behavior with concepts from the health belief model and protection motivation theory. It identifies different stages in the individual appraisal of health messages:

1. People must *realize* that unhealthy eating causes illnesses.
2. People must *acknowledge* that this relationship is significant and that many people suffer from diet-related diseases.
3. People must recognize that they are *personally vulnerable* to this risk.

↓	↓
4. Decided not to act	5. Decided to act
	6. Acting
	7. Maintenance

The opportunities and barriers that the innovative approaches to personalized advice provide for each stage of the behavior change process will be discussed.

Stage 1: Realizing that food influences health

In the first stage, individuals start from a position of being unaware of the health risks of poor food choices. This can be either because the risks are generally unknown or because of *personal ignorance*. When people first learn about the relationship between food intake and health, they are obviously no longer unaware. But although most people are exposed to numerous messages about healthy eating every day, exposure and awareness do not always elicit attention. Through the process of *selective perception* (Sears and Freedman, 1971), people tend to select information that is consistent with their personal attitudes or opinions. Through *cognitive dissonance* (Festinger, 1957), people often ignore information that contradicts their existing beliefs or opinions.

At present, growing Internet use allows for larger access to computer-tailored dietary interventions. However, DeNooijer et al. (2005) note the difficulties in motivating consumers actually to use such interventions, both in 'real world' and study situations. The inclusion of genetic knowledge into personalized nutrition interventions might attract consumer interest. In a recent US market survey, 42% of respondents had heard or read about using individual genetic information for nutrition and diet-related recommendations (Schmidt et al., 2007). Goddard et al. (2007) found a much smaller percentage: only 14% of respondents in the national HealthStyle survey were aware of the availability of nutrigenetic tests offered directly to consumers. Although some people have heard of the availability of tests, this does not indicate their interest in obtaining nutrigenetic testing or their beliefs in the value of such testing. It could be argued that cognitive dissonance can occur among people who hold deterministic beliefs about genes. Schmidt et al. found that more people believed that family history plays a role in health in 2005 (90%) than in 1998 (85%). They argue that this indicates a growing awareness about the interaction between food, genes and health. Yet this awareness does not necessarily lead to an individual motivation to undergo genetic testing and follow nutrition advice personalized to one's genome.

Stage 2: Realizing the significance of healthy eating

In the second stage, people must acknowledge that unhealthy eating impacts health (Rogers, 1983; Janz and Becker, 1984; Ajzen and Madden, 1986), both in physical and social consequences of ill health. In nutrition messages, consequences of conditions like diabetes and cardiovascular disease are most often only explained in terms of *physical* consequences for the individual with the disease. Yet, the social consequences could also substantially impact their everyday life. For instance, the strict medication adherence that is required in diabetes care might interfere with joining sports events or an evening out with friends, but such social consequences are rarely integrated in health messages.

Providing concrete messages about the severity of physical consequences of unhealthy eating is complicated by uncertainties inherent in studying the complex interactions between food and human health, often resulting in equivocal messages why (not) to eat specific foods. For instance, people are confronted with messages that promote the cardiovascular health benefits of olive oil and, at the same time, they are told to reduce their caloric intake because they risk becoming obese.

New knowledge from nutrigenomics research could support development of more concrete messages for healthy eating. Until recently, only genetic diseases such as phenylketonuria and familial hypercholesterolemia have been treated directly through specific dietary intervention, combined with medication in the latter case. But it is likely that nutrigenomics research will lead to more concrete generic messages with respect to complex, common diseases; for instance, that a high intake of omega-3 fatty acids *decreases* the risk of heart disease instead of current messages that omega-3 fatty acids *might lower* the development of heart disease.

Stage 3: Recognizing personal vulnerability

Some currently available nutrition interventions induce awareness of the existence and significance of unhealthy eating (e.g. Van Dillen et al., 2004). But people will only consider behavior change if they also recognize that the information is personally relevant, which means acknowledging that their food intake is not consistent with healthy eating guidelines and makes them vulnerable to diet-related illnesses.

Two issues interfere with recognizing personal vulnerability. First, many people do not know exactly what they eat in comparison to healthy eating guidelines (Lechner et al., 1998, 2006; Oenema and Brug, 2003). One study (Glanz et al., 1997) showed that a substantial portion of adults in the Netherlands and in the USA lacked accurate awareness about their fat consumption. Those people, who inaccurately perceived their own food choice as healthy will have no motivation to change behavior. Second, people use diverse strategies to cope with information about their personal health risk:

- *Defense motivation*: A health threat can induce two coping strategies: it either induces intensive information processing or it induces *defense motivation* (Gleicher and Petty, 1992; Liberman and Chaiken, 1992). The latter is likely to occur when a threat is both severe and challenges personal beliefs. With a defensive motivation, people aim to confirm the validity of their own attitude ('I am eating healthily') and to disconfirm the validity of others ('Your food choices place you at risk'). *Individual biases about personal risk* also influence the perceived threat of unhealthy eating. People tend to overestimate small probabilities with a dramatic impact, such as an airplane crash, and underestimate large probabilities with a more long-term and less dramatic impact, such as heart disease (Koelen and Lyklema, 2004).
- *Unrealistic optimism*: Although people are aware of relative risks of specific behavior, they can have an *unrealistic optimism* towards personal risk (Weinstein, 1980). For instance, people who smoke know that smoking is associated with cancer, but they do not believe they are personally at risk. Van der Pligt (1996) describes several causes underlying unrealistic optimism:
  1. risks that are perceived as *under personal control* induce feelings of optimism
  2. people generally know more about their own protective behavior than about others' behavior; this *egocentric bias* leads to optimism such that people focus more on their own risk-*re*ducing behavior than their risk-*in*ducing behavior
  3. people can have a relatively extreme image of high-risks groups, a *stereo- or prototypical judgment* that does not fit their self-image, leading to optimism

4. people can have a *self-esteem maintenance* mechanism; they generally rate their own actions, lifestyle and personality as better than that of others
5. denial of personal vulnerability is a coping strategy people use to reduce emotional distress, but it undermines the likelihood of preventive actions.

People may use all these mechanisms when confronted with messages about the consequences of unhealthy eating. Their feelings of invulnerability attenuate the perceived personal relevance of the information.

Personalized nutrition interventions aim to tackle the issue of inaccurate perceptions of food choice by providing feedback on current food behavior compared to healthy eating guidelines (Brug et al., 2003). Results of this kind of self-test could also 'correct' users' unrealistic bias about their personal vulnerability by blocking most of the strategies that allow a 'way out'.

Genetic test results can be added to feedback given to people and can serve as a *cue to action*, jointly with the other indicators of personal risk, that is required to become fully aware of one's eating habits. Some research has shown that genetic tests offering great certainty of result, with available treatment and prevention options, are more readily undertaken (Marteau and Croyle, 1998). In contrast, nutrigenomic tests assess the *probability* of developing diet-related illnesses. It is not known whether test results induce defense mechanisms. Given the common perception that genetic risks are immutable, it can be argued that test results induce feelings of fatalism: 'it's in my genes, so what can I do?' (Bouwman et al., 2007).

Stage 4 or 5: Deciding (not) to act

When people consider healthy eating as relevant to them – for example, after they receive personalized nutrition advice from a dietician or an Internet resource – they will consider following recommended nutritional advice. According to Sutton's (1982) extensive review, people evaluate whether the advice will reduce their health risks and the likely physical, mental, social and economic consequences of following healthy eating recommendations. People also take into account whether they are capable of carrying out the advice in their eating practices. This is known as perceived self-efficacy and originates from Bandura's (1982) social learning theory (later called 'social cognitive theory'). Several processes influence self-efficacy, including direct experience, anticipation of consequences and goal-setting. Self-efficacy is the perception of one's own capacity successfully to organize and implement healthy eating largely based on experience with similar actions and situations encountered or observed in the past.

ICT-based personalized interventions aim to influence this decision process by providing feedback tailored to individual characteristics, psychosocial factors, educational level and information needs, making the feedback more personally relevant. First, a user's cognitive state of mind towards changing their food choice is mapped. This is done by means of questionnaires or rating scales that assess psychosocial factors such as attitude, beliefs and perceived self-efficacy towards healthy eating. Second, algorithms are used to find corresponding feedback that facilitates the desired change of those factors. For example, a user with a low perceived efficacy towards healthy cooking will receive easy recipes with step-by-step cooking instructions. Or, a user who believes healthy eating will seriously diminish the taste of meals will receive narratives from a professional cook who talks about healthy, tasty food. The assumption is that this personalized feedback will turn 'barriers' (low perceived self-efficacy) into opportunities (high perceived self-efficacy) and lead to healthy eating.

The influence of genetic test results on decisions to adopt healthy eating advice is scarcely explored. Marteau et al. (2004) found that people who received information about the risk of familial heart disease, including genetic test results, were more likely to perceive their condition as being caused by genes. That perception lowered the expectation that a behavioral means (e.g. eating a low fat diet) would mitigate disease risk and increased the expectation that a biological means (e.g. taking lipid lowering medication) would be effective.

Considering that perceived consequences and perceived self-efficacy strongly influence decisions to act, genetic test results may influence decisions to act in one of two ways: beliefs about the ability to impact health through food choice could be weakened by a deterministic view towards genes and health, *or* beliefs about ability to influence one's own health could be strengthened because the advice is more concrete in terms of its effect on reducing disease risk. At present, it is not known how people will use genetic information and whether it will influence behavior change beyond the information currently supplied, which may take family disease history into account (Haga et al., 2003; Marteau and Weinman, 2006).

Stages 6 and 7: Healthy eating

People who consider healthy eating important tend actively to search for information about healthy eating as the topic becomes more salient to them. They also more frequently discuss the topic with family, friends and health professionals and perhaps already try to cook and eat healthier meals (Blalock and DeVellis, 1998; Lambert and Loiselle, 2007). These activities facilitate people's search for guidance to help them adopt healthier eating routines. Guidance that is specifically tailored to the context of everyday food choice is most likely to aid such behavioral change (Brug et al., 2003; Ayala, 2006).

But changing eating behavior is difficult. Although consumer surveys show that an increasing number of people say they intend to make healthier food choices (Eurobarometer, 2006), a recent food consumption study shows that Dutch people eat too many products that contain saturated and trans fatty acids, while the consumption of fish, fruit and vegetables is too low (Ocke and Hulshof, 2006). US surveys reveal that a majority of the population does not meet national recommendations for vegetable and fruit consumption (Centers for Disease Control and Prevention, 2007). On a global level, the World Health Organization (2004) indicated that people consume too many energy dense, nutrient poor foods that are high in fat, sugar and salt and that people consume too little fruit, vegetables, whole grains and nuts. Increasing rates of obesity and type 2 diabetes highlights this gap between the intention to eat a healthy diet and actual behavior.

Behavioral scientists and anthropologists have argued this gap is caused by a lack of attention to the social and cultural context of food choice (see e.g. Lupton, 1996; Kreuter et al., 2000; Smith, 2004; Brug et al., 2005). The dominant 'nutritionist' perspective focuses on attaining physical health by selecting food products based on their fat, sugar or vitamin content and this perspective guides both research and most nutritional interventions (Scrinis, 2008). Furthermore, a parallel can be drawn between nutritional research and behavioral food research, the research areas that provide the scientific basis for personalized nutrition interventions. Both research areas study how interactions between humans and their social and cultural contexts impact physical health. The areas also share the difficulties involved in exploring contextual variables that often cannot be controlled in research studies (Ajzen, 1992; Fischer, 2006).

If humans are studied without considering contextual influences, the validity of the research results for everyday life situations is limited. This applies to nutritional research, where issues about contextual influences are threefold:

1. limitations of studying single nutrients while people consume food products
2. studying specific food products while people consume diets composed of many foods
3. studying diets without considering the other lifestyle components.

As journalist Michael Pollan (2008) suggests, this perspective causes a gap between healthy eating recommendations (e.g. eat polyunsaturated fats and avoid sugar and saturated fat) and concrete action rules for real life eating practices.

Behavioral research acknowledges that contextual influences, such as the availability and affordability of healthy foods, influence healthy eating. But little is known about the dynamics between an individual's healthy eating intentions and those contextual influences. In the next section, we elaborate on a new approach that takes account of contextual influences to address reasons why many current nutrition interventions are not very successful in inducing healthy eating practices.

THE ACTION APPROACH TOWARDS HEALTHY EATING

The action approach starts from a few considerations. First, it assumes that the context of nutrition behavior is not a set of static factors, but a dynamic situation in which individuals act and react to changing influences. Second, nutrition behavior has two components: it occurs alongside practices of buying food, preparing meals and consuming meals and it is also a discursive practice. People talk with each other about what to buy in the supermarket, what and how meals have to be prepared and how meals are organized, in time and individually and socially. Third, this practice is interwoven with other practices, including child-rearing, work and recreation activities that all interact with one another. For instance, attempting to persuade children to eat vegetables is unavoidably influenced by a certain style in which one attempts to influence their habits generally. To take another case, the way meals are enjoyed on a regular basis (or not) depends on time spent engaged in other activities, such as viewing television, working or sports and other hobbies. Consequently, changing eating habits usually means changing other habits as well and often involves a considerable amount of discursive work. A person who wishes to change eating habits may have to convince others in a family to change food purchasing and consumption choices and has to negotiate eating in social situations where cultural practices often dictate behavior around offering and consuming food. A person may also have to convince themself to control eating practices (e.g. eating only when truly hungry).

To summarize, the action approach does not only address the assessment of the health problem (A), nor the desirable solution, in terms of healthier behavior (B) but concentrates particularly on the trajectory from A to B, taking into account the whole situation in which the behavior is embedded and what is needed to change practices in a desirable direction. Consequently, the action approach envisages the process of creating healthier choices, encompassing all the relevant aspects of the situation.

THE ACTION APPROACH APPLIED TO INNOVATIVE PERSONALIZED NUTRITION INTERVENTIONS

The innovative approaches of using ICT and integrating genetic knowledge can facilitate personalized nutrition interventions. But, as discussed, those innovations do not fully address the challenges people face when

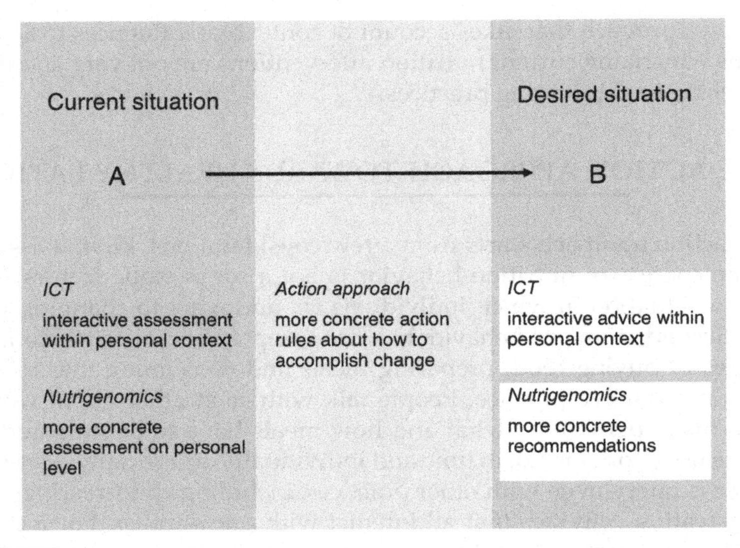

FIGURE 7.1 The contribution of innovative approaches to personalized nutrition advice. A = current situation: food behavior not in line with recommendations about healthy eating. B = desired situation: food behavior in line with recommendations about healthy eating.

they intend to eat healthily in the context of daily life. Those contextual challenges can be addressed by integrating the action approach in the design of personalized interventions, as illustrated by Figure 7.1. In this section, we elaborate on the application of the action approach to ICT-based personalized interventions that incorporate genetic information about disease susceptibilities.

ICT based personalized nutrition intervention

The assumptions of the action approach have several implications for the development of personalized interventions, illustrated in Figure 7.2. First, research must explore the dynamics of healthy eating intentions in practical activities such as buying, preparing and consuming meals, and in discursive practices around eating and in other daily life practices (Figure 7.2: 1). The dynamics will shed light on the challenges and opportunities that people have to deal with when they try to pursue their intentions in daily life situations (Figure 7.2: 2). At present, little is known about these dynamics. In our consumer study, we found that healthy eating intentions were not only undermined by easy accessibility of less healthy choices, but also by cultural norms about how to behave as a dinner guest and the desire to establish oneself as a social person (Bouwman and van Woerkum, in press). ICT applications, such as virtual reality games that mimic eating

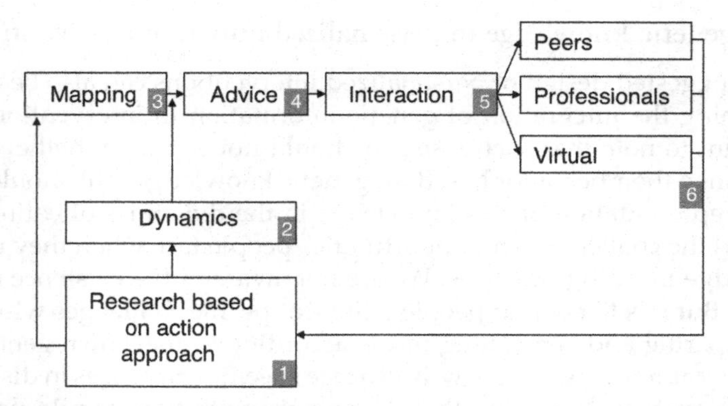

FIGURE 7.2 The co-creation of personalized healthy eating advice, a reflective learning process.

practices, could be used in research to explore dynamics among large study groups. The study results can be used for the development of assessment tools that map the current situation of the user (Figure 7.2: 3) as well as for the development of action rules or guidelines that people can apply in daily life situations (Figure 7.2: 4).

The second implication of the action approach is that the multifaceted nature of food choice complicates the assessment of all dynamics that occur in daily life. Personalized action rules therefore have to be accompanied by interactive tools that mimic those dynamics (see Figure 7.2: 5). For instance, discussion forums with people who received similar or opposing advice, or with health professionals, can facilitate a reflective learning process about how to change eating practices. In addition, interactive, virtual reality applications can prepare people for the dynamics of real-life practices (Bouwman et al., 2005). The additional insights that are derived from those interactions can be added to the available knowledge about dynamics of healthy eating (see Figure 7.2: 6). The third implication of the action approach is that this approach can also be used to attract people's attention. As discussed above, messages have to be consistent with personal beliefs or opinions. Because the action approach studies beliefs in daily life practices, it is likely that messages based on those insights attract more attention than current messages reinforcing 'nutritionism'.

The fourth implication also relates to the impact. Next to reaching a sufficient number of people, interventions have to be effective in changing behavior. At present, most interventions are evaluated based on their effect on actual food intake (e.g. a reduction in fat intake) and on psychosocial factors (e.g. intention to eat a healthy diet). Consistent with the action approach, an evaluation that measures the effect of action rules on the management of challenges in diverse eating practices should be added.

Using genetic knowledge in personalized nutrition interventions

The suggested design of personalized interventions can also be used in researching the integration of genetic information in interventions. It is important to note that such research should not explore whether people will change their behavior based on genetic knowledge, but should focus on the representation of this knowledge in the dynamics of eating practices and the challenges and opportunities people face when they use this knowledge in eating practices. We are not aware of the existence of such studies. But it is likely that people will face specific challenges while buying, preparing and consuming meals according to their own gene-based dietary requirements. People will also face specific challenges in discursive practices such as discussing their test results with their family doctor or other health care provider, especially because recent research indicates that health professionals have a sceptical attitude towards such testing (Bouwman and Te Molder, 2009). Discussing a gene-based diet with a friend who has a deterministic view about the role of genes in maintaining health could further complicate the trajectory from current to desired eating practices.

FINAL CONSIDERATIONS

Innovative personalization approaches in nutritional and behavioral science have the potential significantly to improve the impact of nutrition advice. First, developments in interactive computer technology allow for a sophisticated, personalized assessment of biomedical and behavioral food choice indicators in tailored interventions. Second, nutrigenomics research will allow for advice about nutritional requirements on a more specific level compared to current, generic recommendations. But although promising, those developments will only lead to healthier eating practices if accompanied by the action approach. By taking this approach, people will not only receive personal advice on what they need to change to eat a healthier diet, but also advice on how to accomplish these changes in the context of daily eating practices.

References

Ajzen, I. (1992). Persuasive communication theory in social psychology: a historical perspective. In *Influencing human behavior: theory and applications in recreation and tourism* (M.J. Manfredo, ed.). Sagamore Publishing, Champaign, pp. 1–27.

Ajzen, I. and Madden, T.J. (1986). Prediction of goal directed behaviour: attitudes, intentions and perceived behavioural control. *J Exp Soc Psychol* 22:453–74.

Ayala, G.X. (2006). An experimental evaluation of a group- versus computer-based intervention to improve food portion size estimation skills. *Hlth Educ Res* 21:133–45.

Bandura, A. (1982). Self-efficacy mechanism in human agency. *Am Psychol* 37:122–47.

Blalock, S.J. and DeVellis, R.F. (1998). Health salience: reclaiming a concept from the lost and found. *Hlth Educ Res* 13:399–406.

Bouwman, L., et al. (2005). Personalized nutrition communication through ICT application: how to overcome the gap between potential effectiveness and reality. *Eur J Clin Nutr* 59:108–16.

Bouwman, L., et al. (2007). The personal factor in nutrition communication. In *Personalised nutrition: principles and applications* (F. Kok, ed.). CRC Press, Boca Raton, pp. 169–83.

Bouwman, L. and van Woerkum, C. For health & pleasure. Results of an explorative study among Dutch consumers about the meaning of health in eating. In Press.

Brug, J., et al. (1998). The impact of computer-tailored feedback and iterative feedback on fat, fruit, and vegetable intake. *Hlth Educ Behav* 25:517–31.

Brug, J., et al. (2003). Past, present, and future of computer-tailored nutrition education. *Am J Clin Nutr* 77:1028S–34.

Brug, J., et al. (2005). The internet and nutrition education: challenges and opportunities. *Eur J Clin Nutr* 59:S130–S139.

Centers for Disease Control and Prevention (2007). Fruit and vegetable consumption among adults United States, 2005; CDS; retrieved from http://www.cdc.gov/mmwr/preview/mmwrhtml/mm5610a2.htm; last accessed November 2008.

Curry, S., et al. (2005). A randomized trial of self-help materials, personalized feedback, and telephone counselling with non-volunteer smokers. *J Consult Clin Psychol* 63:1005–14.

DeBusk, R. and Joffe, Y. (2006). *It's not just your genes.* BKDR Publishing, San Diego.

DeNooijer, A. et al. (2005).Bevordering van gezond gedrag via Internet, nu en in de toekomst. ZonMw, Maastricht.

Eng, T. (2004). Population health technologies; emerging innovations for the health of the public. *Am J Prev Med* 26:237–42.

Eurobarometer (ed.) (2006). Health and food. In Special Eurobarometer 246/Wave 64.3-TNS Opinion & Social. European Commission Health and Consumer Protection Directorate General, Brussels.

Festinger, L. (1957). *A theory of cognitive dissonance.* Stanford University Press, Stanford.

Fischer, A. (2006). Social context inside and outside the social psychology lab. In *Bridging social psychology; benefits from transdisciplinary approaches* (P. Van Lange, ed.). Psychology Press, Mahwah.

Glanz, K., et al. (1997). Are awareness of dietary fat intake and actual fat consumption associated? A Dutch-American comparison. *Eur J Clin Nutr* 51:542–47.

Gleicher, F. and Petty, R. (1992). Expectations of reassurance influence the nature of fear-stimulated attitude change. *J Exp Soc Psychol* 28:86–100.

Goddard, K.A.K.A.B., et al. (2007). Awareness and use of direct-to-consumer nutrigenomic tests, United States, 2006. *Genet Med* 9:510–17.

Haga, S.B., et al. (2003). Genomic profiling to promote a healthy lifestyle: not ready for prime time. *Nat Genet* 34:347–50.

Janz, N. and Becker, M. (1984). The health belief model: a decade later. *Hlth Educ Q* 11:1–47.

Koelen, M. and Lyklema, S. (2004). De consument en perceptie van voedselveiligheid. In Ons eten gemeten (C. Kreijl and K. Agac, eds), pp. 241–256. Bohn, Stafleu, Van Loghum, Bilthoven.

Kreijl, C., et al. (2006). *Our food, our health. Healthy diet and safe food in the Netherlands.* RIVM, Bilthoven, pp. 55–92.

Kreuter, M.W. and Stretcher, V.J. (1996). Do tailored behaviour change messages enhance the effectiveness of health risk appraisals? Results from a randomized trial. *Hlth Educ Res* 11:97–105.

Kreuter, M.W., et al. (2000). Are tailored health education materials always more effective than non-tailored materials? *Hlth Educ Res* 15:305–15

Kroeze, W., et al. (2006). A systematic review of randomized trials on the effectiveness of computer-tailored education on physical activity and dietary behaviors. *Ann Behav Med* 31:205–23.

Lambert, S.D. and Loiselle, C.G. (2007). Health information seeking behavior. *Qual Hlth Res* 17:1006–19.

Lechner, L., et al. (1998). Stages of change for fruit, vegetable and fat intake: consequences of misconception. *Hlth Educ Res* 13:1–11.

Lechner, L., et al. (2006). Factors related to misperception of physical activity in The Netherlands and implications for health promotion programmes. *Hlth Promot Int* 21:104–12.

Liberman, A. and Chaiken, S. (1992). Defensive processing of personally relevant health messages. *Pers Soc Psychol Bull* 18:669–79.

Lupton, D. (1996). *Food, the body and the self.* Sage Ltd, London.

Marteau, T.M. and Croyle, R.T. (1998). The new genetics: psychological responses to genetic testing. *Br Med J* 316:693–96.

Marteau, T.M. and Weinman, J. (2006). Self-regulation and the behavioural response to DNA risk information: a theoretical analysis and framework for future research. *Soc Sci Med* 62:1360–8.

Marteau, T., et al. (2004). Psychological impact of genetic testing for familial hypercholesterolaemia in a previously aware population: a randomised controlled trial. *Am J Med Genet* 128A:285–93.

Ocke, M., Hulshof, K., et al. (2006). Food consumption and the intake of nutrients. In *Our food, our health. Healthy diet and safe food in the Netherlands* (C. Kreijl, ed.). RIVM, Bilthoven, pp. 66–75.

Oenema, A. and Brug, J. (2003). Feedback strategies to raise awareness of personal dietary intake: results of a randomized controlled trial. *Prev Med* 36:429–39.

Pollan, M. (2008). *In defense of food; an eater's manifesto.* The Penguin Press, New York.

Rogers, R. (1983). Cognitive and psychological processes in fear appeals and attitude change: A revised theory of protection motivation. In *Social psychology: a source book* (P. Cacioppo, ed.). Guilford Press, New York, pp. 153–76.

Schmidt, D., et al. (2007). US consumer attitudes toward personalized nutrition. In *Personalized nutrition, principles and applications* (F. Kok, ed.). CRC Press, Boca Raton, pp. 205–19.

Scrinis, G. (2008). On the ideology of nutritionism. *Gastronomica* 8:39–48.

Sears, D. and Freedman, J. (1971). Selective exposure to information: a critical review. In *The process and effects of mass communication* (W. Schramm and D. Roberts, eds.). University of Illinois Press, Urbana, p. 209.

Smith, J.L. (2004). Food, health and psychology: competing recipes for research and understanding. *J Hlth Psychol* 9:483–96.

Sutton, S (1982). Fear arousing communications: a critical examination of theory and research. In: *Social Psychology and Behavioural Medicine* (J. Eiser, ed.). John Wiley & Sons, Chichester, p. 303–37

Van der Pligt, J. (1996). Risk perception and self-protective behaviour. *Eur Psychol* 1:34–42.

Van Dillen, S., et al. (2004). Perceived relevance and information needs regarding food topics and preferred information sources among Dutch adults: results of a quantitative consumer. *Eur J Clin Nutr* 58:1306–13.

Weinstein, N. (1980). Unrealistic optimism about future life events. *J Pers Soc Psychol* 39(5):806–20.

Weinstein, N. (1988). The precaution adoption process. *Hlth Psychol* 7:355–86.

WHO (2004). Global strategy on diet, physical activity and health, vol. 2007. World Health Organisation, Geneva.

Further reading

Bouwman, L., et al. (2008) Patients, evidence and genes; an exploration of GPs perspective on gene-based personalized nutrition advice. *Family Practice*, Advance Access, published October 7th, doi: 10.1093/fampra/cmn067.

Bouwman, L. and Te Molder, H. (2008). About evidence-based and beyond: a discourse-analytic study on stakeholders' talk on involvement in the early development of personalized nutrition. *Hlth Educ Res*, Advance Access published on May 21, 2008; doi: 10.1093/her/cyn016.

CHAPTER 8

Health Care Provider Capacity in Nutrition and Genetics – A Canadian Case Study

Jennifer Farrell

OUTLINE

Nutrition and Genomics
ISBN: 978-0-12-374125-7

SUMMARY

Completion of the sequencing of the Human Genome Project in 2003 ushered in a new era of medicine and creates an imperative for health care practitioners to understand the interaction between genetics and the environment, including nutrition, in order to counsel patients on these issues. The question now arises: are health care providers prepared to address genetics and nutrition, including the relationship between the two in nutritional genomics, in their professional practice? This chapter uses a three-part framework of ability, opportunity and motivation to explore issues of health care provider capacity in nutrigenomics, using the Canadian context as a case example. The chapter discusses the education and reported knowledge of physicians, dietitians, pharmacists and naturopaths in the areas of nutrition and genetics, sensitivity to ethical and legal issues, and limitations on current practice based on scope of practice. Public interest in the predictive power of genetics and consumer/patient pursuit of health through nutrition will place increasing demands on health care providers for counseling regarding nutrigenomics and access to nutrigenetic testing services. These professionals ought to be prepared to respond to these demands by gaining appropriate competence in nutrigenomics and/or recognizing where referral to other qualified practitioners is necessary.

INTRODUCTION

As the Greek philosopher Heraclitus observed, one cannot step into the same river twice, for all is in flux. Change in the form of significant discoveries has permeated every field of scientific endeavor, most recently the sequencing of the human genome, which has the potential to transform health care delivery. It has been commented that: '[t]he environment of medicine is changing, maybe slowly and sometimes imperceptibly, but the change will be radical' (Korf, 2005). In the shadow of the Human Genome Project's irresistible progress, however, another issue is looming. The uptake of nutrigenomic science – and genomic science more generally – into clinical practice is hindered by health care providers who are ill-equipped to counsel their patients. Indeed, the impact of genetic discoveries on practice has been likened to 'dumping the Encyclopedia Britannica on a ten-year-old' (Mountcastle-Shah and Holtzman, 2000). To navigate successfully through the challenges nutrigenomics presents them, health care providers will require knowledge of nutrition and genetics that surpasses current capacity. Only when they have capacity to counsel patients on the complex interplay of genetics and the environment in the prevention or progression of disease (Engstrom et al., 2005) can the twin promises of nutrigenomics, that of identification and intervention, be realized. In addition to enhanced core competencies, relevant ethical and legal issues need to be understood before nutrigenomics can be appropriately applied in clinical practice. Outside genetic specialists, many health care providers may view genetics as largely confined to rare single-gene or chromosomal disorders with little relevance for the majority of their patients. An understanding of the complex relationship between genes, lifestyle and environmental exposures is, however, replacing the former genetic and non-genetic divide (Guttmacher et al., 2007). With genetics reaching into many areas of medicine, it has been suggested genetics is 'more like the ocean than a patch of land, lapping upon the shores of continents and islands everywhere' (Korf, 2005). The expansion of genetic relevance will impact primary health care providers on two fronts. First, it will require a new approach to patients that is guided by 'genetic glasses' to tailor care most appropriately. Second, genetic specialists already struggling to meet the demand for their services will require support from primary care providers to meet demand for genetic services (Watson et al., 1999). For these reasons, primary health care providers will be increasingly expected to develop competence in genetics (Watson et al., 1999; Guttmacher et al., 2001). As Guttmacher et al. (2001) suggest, clinical genetics has been needed for the health care of some, but knowledge of the interaction of multiple genes and the environment will be applicable to the health care of all.

This chapter addresses issues of health care provider capacity in nutrition and genetics, using the Canadian context as a case example. The education and reported knowledge of four health care provider groups are analyzed: physicians, dietitians, pharmacists and naturopaths. These groups were selected because consumers and patients may access them directly (i.e. a referral is not needed) and it is likely these groups will be expected to be knowledgeable about nutrition and dietary supplementation either to promote health or to prevent or mitigate disease. A comprehensive literature review was undertaken during summer 2007 to identify studies that have investigated education and training of these professional groups in Canada and other jurisdictions. Keyword searches of major databases of biomedical literature, including Embase and Medline, were conducted. Key terms that were searched in various combinations included: 'genetic(s)', 'nutrition', 'training', 'education', 'knowledge', 'physician', 'dietitian', 'nutritionist', 'pharmacist', 'naturopath', 'health profession(al)'. Additionally, websites for medical, pharmacy, dietetics and naturopathic schools and programs in Canada were identified and curriculum content was examined to assess the number and types of courses in areas related to genetics and nutrition.

Health care providers' capacity to counsel patients in the area of nutrigenomics will be examined through a three-pronged framework of ability, opportunity and motivation (Gray, as cited in Truswell, 2000). Key *abilities* include knowledge, both what is known and what is taught, and sensitivity to ethical and legal issues. *Opportunity* refers to factors that may promote patient interest in seeking nutrigenomic advice and services, but also must take account of limitations that may impede integration of nutrigenomics into health care delivery. Finally, *motivation* includes health care provider attitudes toward, and confidence in, their role in the delivery of nutrigenomic services.

ABILITY

Knowledge

As more gene–disease associations are discovered, primary care providers will face growing demands for information, requests for genetic tests and counseling to help patients interpret results from tests purchased directly from companies (Baars et al., 2005a). Health care providers with insufficient knowledge will not be able to identify individuals who should be tested (Keku et al., 2003), select appropriate tests according to clinical and familial considerations, provide accurate risk assessment (Greendale and Pyeritz, 2001), nor understand the impact of lifestyle choices on the development of disease (Debusk et al., 2005). Guttmacher et al. (2007) caution that: '[h]ealth-care professionals who are ignorant of the basic concepts of medical genetics put their patients at risk of not receiving the best available care, and therefore also put themselves at risk of a malpractice suit'.

The National Coalition for Health Professional Education in Genetics (NCHPEG), an American organization committed to promoting health professional education in human genetics, provides guidance concerning the knowledge base health care providers should possess. NCHPEG (2005) proposes that each health care provider should, at a minimum, have the competence to: 'appreciate [the] limitations of his or her genetics expertise; understand the social and psychological implications of genetic services; [and] know how and when to make a referral to a genetics professional'. Yet educating the workforce to this degree is a daunting task. Much of the genetic instruction in medical schools taught more than 5 years ago is either incorrect or in need of drastic revision (Greendale and Pyeritz, 2001). The challenge is even more formidable when one considers that even if all health care providers beginning their training programs now were adequately educated in nutrition and genetics, and many will not be, when they enter the workforce in several years they will comprise only a small proportion of practitioners (Greendale and Pyeritz, 2001). Even if they did comprise the entire health care workforce, educators of today cannot predict either the knowledge or its clinical application that will be required by today's students over the course of their careers (Guttmacher et al., 2007).

For these reasons, a fresh approach to undergraduate and continuing education is required. The discoveries of the Human Genome Project necessitate changes in professional training curricula so that they may be translated into everyday application (Snow, 2001). Nevertheless, if training remains focused on imparting specific scientific content, it will fail to adapt to the changing health care paradigm. Hayflick and Eiff (2002) call for a focus 'on process at least as much as content' to redirect health care providers' cognitive strategies toward a genetic outlook for each patient. Guttmacher et al. (2007) echo this view, noting that while health care providers must learn specific content, this need will be surpassed by the value of learning key underlying concepts and an appreciation for their significance to everyday practice.

Physicians

Evidence linking diet to disease onset and progression continues to mount (US Department of Health and Human Services, 2000; Heimburger and the Intersociety Professional Nutrition Education Consortium, 2000). Physicians are well situated to provide greater attention to nutrigenomic counseling by virtue of their trusted role in health care delivery (Krebs and Primak, 2006). In a report exploring consumer attitudes to genetic testing, more than 70% of people identified their family physician above all other health care providers as their first choice for genetic information and 80% of individuals vested confidence in their physicians for guidance through the stages of the testing process (Genetic testing: a study of consumer attitudes, as cited in US Secretary's Advisory Committee on Genetics, Health, and Society, 2007). With respect to nutrition, a recent survey indicates that 47% of Canadians received information on food and nutrition from their family physician in the past year, although this figure is down from the 57% reported in a 1994 version of the same study (National Institute of Nutrition and the Canadian Food Information Council, 2004). The study also found that family physicians were considered to be the second most credible source of nutrition information (preceded by dietitians) by more than 40% of those surveyed. Despite the likelihood that Canadians will seek nutrigenomic counseling from their family physicians, this avenue may prove to be unavailable. A gap exists between the information patients expect and what physicians are able to provide.

Physician nutrition education in an international context The inadequacy of physician training in nutrition has been criticized for the past 40 years (Pearson et al., 2001). In a survey of practising US physicians, only 63% indicated they had received adequate training in nutrition in the context of chronic illness (Darer et al., 2004). A recent study of US medical students shows that only 22% believed they had received extensive nutrition training in their senior year (Spencer et al., 2006) and further, in a 2005 US medical school graduation questionnaire, over half the respondents rated the time devoted to nutrition instruction as inadequate (Association of American Medical Colleges, 2005). A survey of US medical schools demonstrates that almost every responding school included some form of nutrition education, however, only 30% of those schools required a separate nutrition course (Adams et al., 2006). The study authors claim the USA is producing physicians who are generally not prepared either to provide nutrition counseling to their patients or to make the correct clinical decisions when nutrition-related issues arise.

Physician nutrition education in a Canadian context Unsurprisingly, the Canadian nutrition education landscape is very similar to the USA. Health educators were aware of the significance of nutrition education dating back to the 1960s (Ng and Hargreaves, 1984), but it has historically received little attention at the undergraduate level of medical training (Murphy, 1993). The Canadian Medical Association's *Code of Ethics* (Update 2004) asks practitioners to '[r]ecognize that community, society and the environment are important factors in the health of individual patients'. Nutrition is a key element of the environment in the bridge between genetic predisposition and disease. Yet, a 2003 study of nutritional advice in family practices reported that most Canadian medical schools offer only a few hours of nutrition instruction and this time is typically associated with specific disorders (Rosser, 2003).

An online search of websites of the 17 medical schools in Canada was undertaken to review curriculum information. Curricula of 12 of these schools (Dalhousie University, McGill University, McMaster University, Memorial University of Newfoundland, Northern Ontario School of Medicine, Queen's University, University of Alberta, University of Calgary, University of Manitoba, University of Saskatchewan, University of Toronto, University of Western Ontario) indicate that required courses in nutrition are offered at 11 schools, with three schools providing additional elective courses. In general, students are required to take only one or two nutrition-related courses throughout the entire program. Some courses focus on community health including topics such as healthy weight standards, dietetic counseling and the relation of nutrition to disease. For example, the second year medicine program at Dalhousie University contains a course on population health that includes nutrition and coronary heart disease as a main theme. Other courses at a variety of schools address scientific aspects such as the biochemistry of the metabolic breakdown of food and the use of biochemical knowledge in the investigation and management of human disease. In the medicine program at McMaster University, topics of study within a Nutrition and Metabolism course include 'nutritional biochemistry, physiology and metabolism at the whole body and cellular level' (McMaster University Electives Manual, 2004/2005). The course description states that: 'it is essential that medical students obtain training in this subject in order to prepare them for a role as nutrition educators and therapists for their patient population' (McMaster University Electives Manual, 2004/2005). However, this course is only offered as an elective and does not form part of the core curriculum. In contrast, the core medicine program at Queen's University aims to give students an understanding of the application of nutritional biochemistry to dietary practices across a person's lifespan, with a focus on dietary assessment methods and tools. Also, learning objectives regarding nutrition have been integrated into other components of the program such as (a) Biochemistry – metabolism of carbohydrates, proteins, and lipids and the associated role of vitamins and minerals; (b) Family medicine – roles of dietitians in the health care system; (c) Geriatrics; and (d) Pharmacology – study of food and drug interactions and toxicity of nutrients in high doses.

Physician genetics education in an international context Inadequate training for physicians in genetics is a concern in many countries. Summarizing research on France, Germany, the Netherlands, Sweden and the UK, Harris et al. (2006) concluded that non-genetics specialist physicians were unaware of genetic issues and features, seldom referred patients for genetic counseling and that undergraduate medical training programs in the UK varied in their basic genetics content and rarely involved applied clinical genetics. In the Netherlands, physicians in another study were found to lack the knowledge required for adequate answers to questions patients ask about genetics, genetic testing and new genetic developments (Baars et al., 2005a). A literature review by Emery et al. (1999) demonstrates that general practitioners from several jurisdictions including Scotland, England, the USA, France and Denmark, have limited genetic knowledge and is further evidence that genetic illiteracy is a concern in many jurisdictions.

Physician genetics education in a Canadian context A recent study (Thurston et al., 2007) of genetics instruction in US and Canadian medical schools claims that all responding schools adequately cover the topics of genetic heritability, molecular biology and population genetics and that many also teach the skills of how to take a multi-generational family history, determine pattern inheritance and assess each family member's risk. Further, most schools also educate their students on how to interpret genetic test results and how to convey them to the patient. These findings indicate an improvement over previous reports that showed almost one third of Canadian physicians received no formal training in genetics (Bottorff et al., 2005). The Ontario Report to Premiers (2002) on genetics also claims that all physicians receive basic education in genetic susceptibilities.

A review of online curricula information was conducted for the 17 medical schools in Canada. The websites of 13 of these schools (Dalhousie University, McGill University, McMaster University, Memorial University of Newfoundland, Northern Ontario School of Medicine, Queen's University, University of Alberta, University of British Columbia, University of Calgary, University of Manitoba, University of Saskatchewan, University of Toronto, University of Western Ontario) indicate that required courses in genetics are offered at 11 schools, with two schools providing additional elective courses. One school offers elective courses only. In general, students are required to take only one or two genetics courses throughout the entire program. Overall, course content includes the study of classical and molecular genetics as well as the use of genetics services and counseling. Examples of course objectives that demonstrate a link between genetics and disease include: 'apply principles of medical genetics in clinical problem solving' and 'describe the components of taking a genetic history' (University of Manitoba, 2007). Some evidence indicates that very few medical students, as little as 1–3%, take courses in genetics when they are offered as electives (Baars et al., 2005b).

Link between nutrition and genetics in physician training programs Our review did not find medical school courses that specifically address nutritional genomics. Pharmacogenomics is addressed in some curricula (see e.g. Medical Genetics curriculum at Queen's University).

Dietitians

In some jurisdictions, especially those where physician services are publicly insured and physicians enjoy a high degree of public trust and respect, consumers may principally seek nutrigenomics information and advice from physicians. Practice workload and time constraints will, however, necessitate alliances with professionals such as dietitians (Christianson et al., 2005). A study assessing Canadians' level of confidence in their sources of nutrition information indicates that 88% are very confident about information provided by dietitians (Marquis et al., 2005). Dietitians' background in the science and management of diet and nutrition positions them to be key providers of nutrigenomic services (Debusk et al., 2005). They will, however, need to adapt to a new area of practice by incorporating genetics into their competencies (Debusk et al., 2005). Kauwell (2003) suggests that as nutrigenomics expands, dietitians will need to add new tasks to their repertoire such as doing risk assessments based on elicited genetic information, understanding genetic test results, communicating the roles of genetics and diet in disease, offering nutrition intervention strategies, counseling patients on the significance of genetic disorders to family members, achieving informed decision-making and knowing which cases should be referred for genetic testing and/or counseling. Kauwell

(2003) cautions that the credibility of the profession is at stake and explains that dietitians, as skilled counselors and educators, must be ready to face genetic enquiries and concerns and be prepared to respond with at least a basic grasp of genetics, resource sharing and making referrals if necessary.

Dietitians' education in an international context In the interest of brevity, nutrition training of dietitians will not be examined here, as the adequacy of such education is assumed to be sufficient. In contrast, capacity in genetics appears to be lacking. Survey results from US directors of Didactic Programs in Dietetics point to weaknesses in genetic training since current curricula contain little or no genetics content (Vickery and Cotugna, 2005). In another survey of US dietitians, more than one third had no formal genetics education (Gilbride and Camp, 2004). Atkinson (2006) proposes that the modern field of nutrition has splintered into several sub-specialties such as nutrigenomics, nutraceuticals, natural health products, functional foods and many more. She argues that: '[a] general undergraduate degree in nutrition or food science simply does not prepare one for working in these subspecialty areas' (Atkinson, 2006).

Dietitians' education in a Canadian context A review of online curriculum information was conducted for the 16 dietetic programs in Canada. The websites of 10 of these programs (University of Saskatchewan, University of Manitoba, University of Guelph, University of Prince Edward Island, Acadia University, McGill University, Memorial University of Newfoundland, University of Alberta, University of British Columbia, University of Western Ontario) indicate that required courses specific to genetics are offered in five programs, with two programs offering additional elective genetics courses. Two programs offer elective courses only. Overall, most courses cover classical, molecular and population genetics.

Link between nutrition and genetics in dietitian training programs Some courses offer integrated nutrition and genetics content. The following elective course objective is indicative of other similar courses: 'applying knowledge learned in basic nutrition, physiology, genetics, and biochemistry to help develop an understanding of the metabolic bases of the interactions of nutrients under physiological and various pathological states' (University of British Columbia, 2007).

Naturopaths

Naturopath nutrition education in a Canadian context Naturopathic medicine is a mode of primary health care that emphasizes disease prevention as well as correction of the underlying pathology (College of Naturopathic Physicians of British Columbia, 2006). Training is modeled in the same vein as standard medical education. There are two naturopathic medicine schools in Canada: the Canadian College of Naturopathic Medicine (Toronto) and the Boucher Institute of Naturopathic Medicine (Vancouver). One offers four courses in nutrition that cover metabolism; diet quality; special nutritional needs related to pregnancy, nursing, infancy, and vegetarianism; nutrient supplementation; supplement programs relevant to specific diseases and stages of the human life cycle; and food habits. The other school also offers four relevant courses, one in biochemistry that covers the metabolic breakdown of food as it applies to health and disease, and three in clinical nutrition that focus on the macro- and micronutrients required for human health in terms of physiological function, diet and lifestyle intervention plans allowing for primary care management of metabolic diseases, and the toxicology of common vitamins. Clinical nutrition represents one of six major disciplines that define naturopathic medicine and a survey of Canadian naturopaths indicates that 85% claim to have received adequate or thorough coverage of nutritional counseling in their training (Verhoef et al., 2006). Nonetheless, only 25% of Canadians report feeling very confident about nutrition information from naturopaths (Marquis et al., 2005).

Naturopath genetics education in a Canadian context One naturopathic medicine school in Canada offers one course in pathology that covers the genetic basis of disease and one course in genetics that examines the contribution of rapid changes in genetics to the field of medicine. The second school offers a course in pathology but no specific courses in genetics.

Link between nutrition and genetics in naturopath training programs One of the two naturopathic medicine schools in Canada includes a clinical nutrition module which, citing nutrition as the cornerstone of good health, covers the interacting biochemical roles of nutrients in human metabolism, clinical signs of nutritional imbalances, food sensitivities, metabolic disorders and the inborn errors of metabolism including the consideration of genetic factors. Biochemical individuality is a central topic within one of the clinical nutrition courses. This is consistent with one of the principles of naturopathic medicine, namely 'to heal the whole person through individualized treatment' (Canadian College of Naturopathic Medicine, 2007/2008).

Pharmacists

Many advocate for a more interdisciplinary approach to health care delivery in Canada (see e.g. Lett, 2008). A report by Canada's National Drug Scheduling Advisory Committee (2004) suggested that, under a new model of interdisciplinarity, pharmacists will be increasingly relied upon as primary caregivers and will assume a gatekeeper role previously monopolized by physicians. This report proposes that pharmacies will move beyond 'drug stores' to primary health care centers. This evolving role is evidenced by recent legislation conferring prescribing powers on pharmacists in one Canadian province (see Alberta *Health Professions Act* and the *Pharmacy and Drug Act*).

Pharmacists will need a broader knowledge base to meet their expanding professional responsibilities. A new generation of personalized medications coupled with the impact of genotyping on routine prescriptions of common medications will necessitate greater genetic literacy (Ontario Report to Premiers, 2002). Also, the pharmacogenomic concept of individualized medicine is growing to include the field of nutrition (Kaput and Rodriguez, 2004). Pharmacogenomics and nutrigenomics have been described as being linked by a continuum from pharmaceuticals to nutraceuticals (Ghosh et al., 2007). These two scientific fields will become even more connected as the relationships between foods and drugs are more thoroughly understood (Ghosh et al., 2007). Furthermore, if pharmacists gain prescribing power, an understanding of nutrigenomics will be more urgent as food substances with therapeutic properties similar to pharmaceuticals are identified. Pharmacists may also feel pressure from consumers who, having pursued direct-to-consumer nutrigenomic services, seek advice about dietary supplements. In addition, supplements may interact with an individual's drug requirements and consumers will turn to pharmacists for advice as drug experts. At present, some pharmacy curricula in Canada already incorporate natural health product content and some pharmacy professional bodies emphasize the need to counsel patients on such products (Farrell et al., 2008).

A study of Canadian and US pharmacy programs indicates that 87% of pharmacy schools included pharmacogenomics in their curricula and, of these, it was a required element in 95% of the schools (Zdanowicz et al., 2006). Interestingly, more than 70% of the programs with pharmacogenomics content also include information on human genomics, suggesting that pharmacy schools have been proactive in keeping curricula current.

Pharmacist nutrition education in a Canadian context A review of online curricula information was conducted for the 11 pharmacy programs in Canada. The websites of six (Dalhousie University, Université Laval, University of Alberta, University of Manitoba, University of Saskatchewan, University of Toronto) of these programs indicate that required nutrition courses are offered in four programs. Two programs offer only elective courses. In general, students are required to take only one nutrition course throughout the entire program, with only one school requiring students to take two courses. Courses focus on supplements, biochemical process of digestion, absorption and metabolism, nutrition and disease, natural health products, dietary guidelines and nutrition for optimal health.

Pharmacist genetics education in a Canadian context A review of online curricula information was conducted for the 11 pharmacy programs in Canada. The websites of five (Université Laval, University of Alberta, University of Manitoba, University of Saskatchewan, University of Toronto) of these programs indicate that required nutrition courses are offered in four programs. Of these five, no courses specific to genetics are offered, but four schools offer courses that contain some level of genetics content. Two relevant courses, one of which is optional, are related to natural health products, and one other course covers, among other objectives, the human genome and its implication in pharmacy. Finally, genetic content in another course includes legal, ethical and economic issues associated with pharmaceutical biotechnology, the principles of biotechnology as they apply to the development of pharmaceutical products and the uses of these products in the treatment of various conditions.

Summary

Physicians in Canada currently appear to receive mandatory training in both nutrition and genetics. In contrast, dietitians, naturopaths and pharmacists each receive nutrition training, although their exposure to genetics training is minimal. Further, among all four disciplines, courses that link nutrition and genetics content are rare.

Ethical and Legal Awareness

Ethical and legal concerns in nutrigenomics involve tensions between rights and duties and health professionals face liability exposure if they practice without due attention to these issues. In practical terms, this calls for mandatory ethical-legal training so providers will be able both to recognize and manage ethical and legal aspects of nutrigenomics applications. Ethics education has increased in recent years. A 2004 survey of medical ethics education at US and Canadian medical schools indicates that all responding schools, including six Canadian, provided formal instruction in medical ethics (Lehmann et al., 2004). This is a positive trend as some predict that the future of medical genetics depends on health care providers' ability to incorporate genetic advances into their practices in accordance with governing ethical principles (Hook et al., 2004).

Issues of confidentiality, privacy and informed consent are important in nutrigenomics service delivery. These are highly protected and valued rights in health care and are linked together through the common threads of autonomy and personal integrity. To some, genetic information is unlike other health information. It is argued that genetic information can be set apart by virtue of its personal nature, risks of discrimination, its potentially predictive power, its application across the lifespan and its pertinence to biologic family members (Noonan, 2002). In contrast, others argue that non-genetic information also shares some of these characteristics and genetic information can be protected within existing ethical and legal frameworks. For example, tests for infectious disease may implicate family members and diagnosis of mental illness can stigmatize the affected individual. To the extent that genetic information is perceived as deserving different treatment, ethical and legal issues will take on a new level of complexity.

Confidentiality refers to the duty of a health care provider to safeguard patients' personal information from unauthorized use or disclosure. Confidentiality is vital to the trust relationship between provider and patient and is a binding obligation entrenched in professional norms and codes of ethics, and by various statutes, regulations and common law principles (Lacroix et al., 2005).

Closely related to confidentiality is the right to privacy, which is the expectation of personal concealment from the scrutiny of others (Ellerin et al., 2005). One aspect of privacy is the right not to know genetic information. Genetic information is one of the most personal and sensitive types of information, indeed of the same nature as private thoughts, fantasies and dreams (Shaw, 1987) and learning it can dramatically alter a person's self-perception and identity. Laurie (1996) argues that true autonomy must incorporate a choice not to learn of personal information and that to impose unwanted information onto a patient amounts to an invasion of privacy. To illustrate this point, Laurie cites philosopher James Fitzjames Stephen, who wrote that: '[p]rivacy may be violated not only by the intrusion of a stranger, but by compelling or persuading a person to direct too much attention to his own feelings and to attach too much importance to their analysis' (1996). Health care providers should be aware that some patients may express a desire not to know genetic information and, in counseling, provide sufficient information to allow patients to make an informed choice about testing.

Tensions involved in privacy and confidentiality may arise in familial relationships. The inherent nature of genetic information means that, to differing degrees of certainty, information about biological family members is revealed (Laurie, 1996). While most recipients of genetic risk information communicate it to family members due to the potential benefit of risk awareness, some individuals are unwilling to do so (Lacroix et al., 2005). To the extent that the risk is for a serious disorder and steps to minimize risk or progression are possible, a tension arises for the health care provider between a duty to protect privacy and maintain confidentiality and a duty to act for the benefit of others (Lacroix et al., 2005). Health professionals' duty of confidentiality is not absolute and they may feel an obligation or desire to inform family members about genetic risk information even without the consent of his or her patient. Non-consensual disclosure of personal information is legally and ethically permissible in some limited circumstances to protect public safety and avert risk of harm to identifiable third parties (Knoppers and Cardinal, as cited in Downie et al., 2002). In a survey of US genetic counselors, 63% believed they have an obligation to communicate genetic information to their patient's relatives even when their patient does not consent, however, when asked about past encounters with a patient refusing to contact at-risk relatives, 24 of 25 counselors did not breach their duty of confidentiality (Dugan et al., 2003).

Informed consent also gains a new level of complexity in the application of genomics to health care practice. Voluntary, informed consent is a legal and ethical prerequisite to treatment and must be provided prior to genetic testing. To provide this consent, a patient must have the capacity to understand the benefits and material risks involved in testing and be informed of the medical procedure. For predictive genetic testing, the patient must understand all aspects of testing including the disorder being tested for, the utility of the test, treatment options and the medical and non-medical consequences of test results (Ontario Report to Premiers, 2002). Yet, the complexity and limited accuracy of some tests, together with the rapid pace of new scientific information, cast doubt on the standard of disclosure required of health care providers to achieve informed consent (Caulfield, 1999; Ontario Report to Premiers, 2002). Further, patients are more likely to experience difficulty in understanding genetic language and concepts of susceptibilities and probabilities. This difficulty is exacerbated if health care providers themselves lack the sophisticated knowledge required to explain the full impact of testing to the patient. In these cases, patients may be making decisions in the absence of true autonomy (Rosas-Blum et al., 2007) and hence fail to give informed consent.

Other legal and ethical issues arise as primary care providers adopt a greater role in genetic service delivery. For instance, at what point along the continuum of genetics service delivery do health care providers hold themselves out as genetics experts, thereby triggering a higher standard of care in the law of negligence? These are issues without clear answers and the complexity of genetic scenarios will likely resist a rigid ethical framework. Rather, professional judgment will play a vital role in reasoning through these issues (Lacroix et al., 2005), representing a further element to building capacity to handle nutrigenomics in practice.

Skills

In addition to the knowledge acquired through an education in nutrition and genetics, health care providers will require sound counseling skills to ensure effective nutrigenomics service delivery. Wang et al. (2004) identify three broad objectives of genetic counseling: first, to educate and inform patients of the genetic disorder or gene–disease association; second, to provide psychological and social support including referral if necessary; and third, to foster informed decision-making. The effectiveness and success of genetic counseling will hinge on the degree to which these goals are reached. Effective communication has been linked to greater patient satisfaction, improved commitment to medical regimens and better treatment response for chronic diseases (Rosas-Blum et al., 2007). Conversely, poor communication is associated with higher health care costs, unnecessary pain and heightened fear and anxiety (Rosas-Blum et al., 2007). A study assessing US physicians' opinions of genetic testing indicated that more than 90% believed testing should be accompanied by counseling regarding the benefits, risks and possible consequences of the test, yet only 29% felt qualified to provide it (Freedman et al., 2003). This finding may be representative of a lack of training since complex material must be thoroughly understood before it can be simplified for patients (Rosas-Blum et al., 2007).

OPPORTUNITY

Scope of Practice

Regulated health care providers must adhere to the boundaries of their legally permissible scope of practice. As one would expect, physicians in Canada have the most inclusive scope of practice among the health professions. Of key relevance to nutrigenomics is the physician's authority to diagnose the patient; to promote wellness, disease prevention and cure; and to prescribe treatment including drug therapy. (For example, see Ontario's *Medicine Act*, which states that the practice of medicine is: 'the assessment of the physical or mental condition of an individual and the diagnosis, treatment and prevention of any disease, disorder or dysfunction', *Medicine Act*, 1991, S.O. 1991, c. 30, s.3.) Dietitians, in contrast, have a very limited scope of practice, permitting activities such as nutritional assessment and the treatment and prevention of nutrition-related disorders by nutritional means (*Dietetics Act*, Ontario, s.3). Dietitians across all provincial jurisdictions lack the authority to diagnose their clients. A guideline from the College of Dietitians of Ontario (2004) explains that diagnosis means to identify to a patient 'the name of a disease or disorder from which the client is suffering in circumstances in which it is reasonably foreseeable that the client/representative will rely on the diagnosis'. Two permissible acts include discussing a formerly communicated diagnosis and informing clients of laboratory results if the client has prior knowledge of their disease condition. While dietitians cannot convey a diagnosis to clients directly, the College of Dietitians of Ontario (2006) notes that their members are permitted to formulate and communicate a diagnosis to other health care providers.

Dietitians in Canada are also not permitted to prescribe drug therapy. One province (Alberta) permits a broader scope of practice for some dietitians who, upon demonstrating the required competence, may be authorized to prescribe specified drugs for the purposes of providing nutritional support (see s.10(1)(c) Registered Dietitians and Registered Nutritionists Profession Regulation, Alta. Reg. 79/2002). With respect to nutrigenomics then, the capacity of dietitians to provide these services would be limited to that which follows genetic testing and interpretation, i.e. to the development and implementation of an appropriate diet exclusive of drugs.

Naturopathic medicine is currently regulated as a health profession in several Canadian provinces and naturopathic doctors may be well situated to incorporate nutrigenomics into practice. They are authorized to diagnose patients through the use of, among other tools, physical examinations and comprehensive laboratory testing. Their expertise in natural and preventive medicine, together with their strong education in nutrition and lifestyle counseling, if combined with training in genetics, would position them to incorporate nutrigenomics into their practice.

Finally, nutrigenomics will likely have some impact on pharmacists in Canada. Consumers may seek pharmacists' advice regarding self-testing products and natural health products sold through pharmacies. Self-tests are devices approved by the federal government for use by the general public to help individuals self-diagnose a specific condition with symptoms, screen for a condition with symptoms or monitor existing conditions. Examples of self-testing kits that may be provided and explained by a pharmacist include blood pressure, blood glucose monitoring, pulmonary functioning and cholesterol tests. Consumers with genetic predispositions to heart disease or hypercholesterolemia may use these tests and may also seek advice on vitamins, herbal preparations or other natural health products that may be helpful in reducing disease risk.

Workload

Health care providers' heavy workload is another obstacle to incorporating nutrigenomics into their practice. Six thousand new articles are published each week in medical literature and more than 1000 new guidelines are published every year (Rosser, 2003), imposing a heavy burden on practitioners who seek to stay current with new research findings. In a study investigating the barriers to nutrition counseling among primary care providers, lack of time was the most frequently encountered obstacle and was reported by 75% of those surveyed (Kushner, 1995). In a separate study that examined physician involvement in genetic counseling, one participant remarked: '...I've no idea how long a genetic history would take but if you were to add that on to the day I don't know how that would be resourced' (Watson et al., 1999). A second participant commented: '. . .we are (becoming) more involved with everything – it's not just genetics...' (Watson et al., 1999). Clearly, the demands on health care providers' time are an important barrier to providing nutrigenomic services.

MOTIVATION

Attitudes

Emerging genetic medicine is exerting pressure on primary care providers to assimilate more sophisticated services into their responsibilities, but they typically desire only a limited role in the provision of services despite having positive attitudes toward nutrition and genetics (Emery et al., 1999; van Weel, 2003). Fry et al. (1999) found many primary care providers are strongly opposed to taking on an increasingly specialist role, preferring instead familiar activities such as being a gatekeeper to specialist services, providing counseling and taking family histories. Yet, even these customary roles will take on new meaning. In the role of gatekeeper for example, the primary care provider must be able to identify those at heightened risk of disease and to clarify complex advice communicated to the patient by the genetic specialist. In an era of genetic medicine, performing standard tasks with a genetic twist will require new knowledge.

Confidence

Primary care providers in Canada lack confidence in some areas of patient care. Bottorff et al. (2005) found that confidence was deficient among physicians even in routine activities such as screening, counseling and giving lifestyle advice. In another study of Canadian physicians, a majority perceived their genetics knowledge to be adequate, but only a minority had sufficient confidence to offer genetic counseling for straightforward genetic scenarios (Hunter et al., 1998). Dietitians also report lack of self-confidence, indeed, one study found that 60% indicated low confidence levels toward nutrigenomics application (Rosen et al., 2006).

Perceptions that genetics is complicated and intimidating may undermine confidence. Shields et al. (2005) report survey data showing that primary care physicians were 11% less likely to offer a test to patients if it was labeled a 'genetic' test as opposed to the same test being labelled a 'serum protein' test.

CONCLUSION

Health care systems and institutions, including institutions that train health care providers, are not amenable to rapid changes in their operations. But these bodies should not wait until the intricacies of the human genome are fully understood before adopting ways to prepare health care providers to incorporate genomics into practice. As Gurwitz et al. (2003) suggest, the essence of academia is to 'teach even what is not entirely understood, in the hope that our students will expand the current knowledge'. One could also argue that if society desires to benefit from genomic advances, then it surely owes health care providers the opportunity to learn the required knowledge and to become familiar with the ethical and legal implications of genetic testing and services. Otherwise, public enthusiasm for the predictive power of genetics and their pursuit of health will demand services that providers are ill-equipped to give. This will lead to premature testing applications and an invitation to ethical and legal mismanagement. Attention to the obstacles discussed in this chapter will ease the implementation of nutrigenomics into practice. Although the promise of nutrigenomics calls for patience and responsible action, the obstacles facing today's students and health care providers are considerable and addressing them must be a pressing matter for educators.

References

Adams, K.M., et al. (2006). Status of nutrition education in medical schools. *Am J Clin Nutr* 83(suppl):941S–44.

Association of American Medical Colleges, Division of Medical Education (2005). Medical School Graduation Questionnaire, Final All Schools Report. Accessed 1 January, 2008, <http://www.aamc.org/data/gq/allschoolsreports/2005.pdf>.

Atkinson, S.A. (2006). A nutrition odyssey: knowledge discovery, translation, and outreach, 2006 Ryley-Jeffs Memorial Lecture. *Can J Diet Pract Res* 67:150–56.

Baars, M.J.H., et al. (2005a). Deficiency of knowledge of genetics and genetic tests among general practitioners, gynecologists, and pediatricians: a global problem. *Genet Med* 7:605–10.

Baars, M.J.H., et al. (2005b). Deficient knowledge of genetics relevant for daily practice among medical students nearing graduation. *Genet Med* 7:295–301.

Bottorff, J.L., et al. (2005). The educational needs and professional roles of Canadian physicians and nurses regarding genetic testing and adult onset hereditary disease. *Commun Gene* 8:80–7.

Canadian College of Naturopathic Medicine, The School of Continuing Education, Course Calendar (2007/2008). Accessed 4 January, 2008, http://www.ccnm.edu/files/pdfs/ce/Cecalendar2006-07_web.pdf.

Canadian Medical Association, 2004. Code of Ethics (Update 2004). Accessed 6 January, 2008, <http://policybase.cma.ca/PolicyPDF/PD04-06.pdf>.

Caulfield, T. (1999). *Gene testing in the biotech century: are physicians ready? Can Med Assoc J* 161:1122–4.

Christianson, C.A., et al. (2005). Assessment of allied health graduates' preparation to integrate genetic knowledge and skills into clinical practice. *J Allied Hlth* 34:138–44.

College of Dietitians of Ontario (2004). Guidelines: controlled acts. Accessed 5 January, 2008, <http://www.cdo.on.ca/en/pdf/publications/guidelines/controlledActs.pdf?printVersion=no>.

College of Dietitians of Ontario (2006). Are you considering a leave of absence? Accessed 5 January, 2008, <http://www.cdo.on.ca/en/pdf/Publications/resume/resumesummer06-Eng.pdf>.

College of Naturopathic Physicians of British Columbia (2006). Prescriptive authority for naturopathic physicians: objectives, rationale and a framework for regulation: a proposal for the British Columbia Ministry of Health, 22 December 2006. Accessed 1 January, 2008, <http://www.cnpbc.bc.ca/PDF-2007/Prescriptive%20Authority %20Proposal.pdf>.

Darer, J.D., et al. (2004). More training needed in chronic care: a survey of US physicians. *Acad Med* 79:541–48.

Debusk, R.M., et al. (2005). Nutritional genomics in practice: where do we begin?. *J Am Diet Assoc* 105:589–98.

Downie, J., et al. (ed.) (2002). *Canadian health law and policy*, 2nd edn. LexisNexis Canada Inc., Toronto.

Dugan, R.B., et al. (2003). Duty to warn at-risk relatives for genetic disease: genetic counselors' clinical experience. *Am J Med Genet* 119A:27–34.

Ellerin, B.E., et al. (2005). Ethical, legal, and social issues related to genomics and cancer research: the impending crisis. *J Am Coll Radiol* 2:919–26.

Emery, J., et al. (1999). A systematic review of the literature exploring the role of primary care in genetic services. *Fam Pract* 16:426–45.

Engstrom, J.L., et al. (2005). Genetic competencies essential for health care professionals in primary care. *J Midwif Women's Hlth* 50:177–83.

Farrell, J., Ries, N. and Boon, H. (2008). Pharmacists and natural health products: a systematic analysis of professional responsibilities in Canada. *Pharm Pract* 6:33–42.

Freedman, A.N., et al. (2003). US physicians' attitudes toward genetic testing for cancer susceptibility. *Am J Med Genet* 120A:63–71.

Fry, A., et al. (1999). GPs' views on their role in cancer genetics services and current practice. *Fam Pract* 16:468–74.

Ghosh, D., et al. (2007). Pharmacogenomics and nutrigenomics: synergies and differences. *Eur J Clin Nutr* 61:567–74.

Gilbride, J.A. and Camp, K. (2004). Preparation and needs for genetics education in dietetics. *Topics Clin Nutr* 19:316–23.

Greendale, K. and Pyeritz, R.E. (2001). Empowering primary care health professionals in medical genetics: How soon? How fast? How far?. *Am J Med Genet* 106:223–32.

Gurwitz, D., et al. (2003). Education: teaching pharmacogenomics to prepare future physicians and researchers for personalized medicine. *Trends Pharmacol Sci* 24:122–25.

Guttmacher, A.E., et al. (2001). Genomic medicine: who will practice it? A call to open arms. *Am J Med Genet* 106:216–22.

Guttmacher, A.E., et al. (2007). Educating health-care professionals about genetics and genomics. *Nature* 8:151–57.

Harris, R., et al. (2006). Genetic education for non-geneticist health professionals. *Commun Genet* 9:224–26.

Hayflick, S.J. and Eiff, P.M. (2002). Will the learners be learned?. *Genet Med* 4:43–4.

Heimburger, D.C. and the Intersociety Professional Nutrition Education Consortium (2000). Physician-nutrition-specialist track: if we build it, will they come? *Am J Clin Nutr* 71:1048–53.

Hook, C.C., et al. (2004). Primer on medical genomics part XIII: ethical and regulatory issues. *Mayo Clin Proc* 79:645–50.

Hunter, A., et al. (1998). Physician knowledge and attitudes towards molecular genetic (DNA) testing of their patients. *Clin Genet* 53:447–55.

Kaput, J. and Rodriguez, R.L. (2004). Nutritional genomics: the next frontier in the postgenomic era. *Physiol Genomics* 16:166–77.

Kauwell, G.P.A. (2003). A genomic approach to dietetic practice: are you ready?. *Topics Clin Nutr* 18:81–91.

Keku, T.O., et al. (2003). Gene testing: what the health professional needs to know. *J Nutr* 133:3754S–7.

Korf, B.R. (2005). Genetics in medical practice: the need for ultimate makeover. *Genet Med* 7:293–94.

Krebs, N.F. and Primak, L.E. (2006). Comprehensive integration of nutrition into medical training. *Am J Clin Nutr* 83(suppl):945S–50.

Kushner, R.F. (1995). Barriers to providing nutrition counseling by physicians: a survey of primary care practitioners. *Prev Med* 24:546–52.

Lacroix, M., et al. (2005). Warning patients' relatives of genetic risks: policy approaches. *GenEditorial* 3:1–8.

Laurie, G.T. (1996). The most personal information of all: an appraisal of genetic privacy in the shadow of the human genome project. *Internat J Law Policy Fam* 10:74–101.

Lehmann, L.S., et al. (2004). A survey of medical ethics education at US and Canadian medical schools. *Acad Med* 79:682–89.

Lett, D. (2008). The new architecture of medical education. *Can Med Assoc J* 178:17.

Marquis, M., et al. (2005). Canadians' level of confidence in their sources of nutrition information. *Can J Diet Pract Res* 66:170–75.

McMaster University (2004/2005). Faculty of Health Sciences Electives Manual. Accessed 20 April, 2008, <http://65.39.131.180/ContentPage.aspx?name=Undergraduate_MD_Program_Elec_Mac>.

Mountcastle-Shah, E. and Holtzman, N.A. (2000). Primary care physicians' perceptions of barriers to genetic testing and their willingness to participate in research. *Am J Med Genet* 94:409–16.

Murphy, P.S. (1993). Nutrition activities of physicians in their family practice setting: changes following a continuing education nutrition program. *J Can Diet Assoc* 54: 208–11.

National Coalition for Health Professional Education in Genetics (2005). The core competencies. Accessed 12 December, 2007, <http://www.nchpeg.org/core/corecomps 2005.pdf>.

National Drug Scheduling Advisory Committee (2004). A preliminary review of the relevance of Canada's national drug scheduling system: a report to the Council of Pharmacy Registrars of Canada, February, 2004. Accessed 3 January, 2008, <http://www.napra.ca/pdfs/drugsched/preliminary_review_NDS_system-final.pdf>.

National Institute of Nutrition and the Canadian Food Information Council (2004). Tracking Nutrition Trends V, Accessed 1 January, 2008, <http://www.ccfn.ca/pdfs/TNTV-FINAL.pdf>.

Ng, M.L. and Hargreaves, J.A. (1984). Status of nutrition education in Canadian dental and medical schools. *Can Med Assoc J* 130:851–53.

Noonan, A.S. (2002). Key roles of government in genomics and proteomics: a public health perspective. *Genet Med* 4(Suppl):72S–6.

Ontario Report to Premiers (2002). Genetics, testing & gene patenting: charting new territory in healthcare, January 2002. Accessed 2 January, 2008, <http://www.health.gov.on.ca/english/public/pub/ministry_reports/geneticsrep02/report_e.pdf>.

Pearson, T.A., et al. (2001). Translation of nutritional sciences into medical education: the Nutrition Academic Award Program. *Am J Clini Nutr* 74:164–70.

Registered Dietitians and Registered Nutritionists Profession Regulation, Alberta Regulation 79/2002.

Rosas-Blum, E., et al. (2007). Communicating genetic information: a difficult challenge for future pediatricians. *BMC Med Educ* 7:17.

Rosen, R., et al. (2006). Continuing education needs of registered dietitians regarding nutrigenomics. *J Am Diet Assoc* 106:1242–5.

Rosser, W.W. (2003). Nutritional advice in Canadian family practice. *Am J Clin Nutr* 77(suppl):1011S–5.
Shaw, M.W. (1987). Invited editorial comment: testing for the Huntington gene: a right to know, a right not to know, or a duty to know. *Am J Med Genet* 26:243–46.
Shields, A.E., et al. (2005). Barriers to translating emerging genetic research on smoking into clinical practice: perspectives of primary care physicians. *J Gen Intern Med* 20:131–38.
Snow, K. (2001). The growing impact of genetics on health care: do we have appropriate educational resources?. *Mayo Clin Proc* 76:769–71.
Spencer, E.H., et al. (2006). Predictors of nutrition counseling behaviors and attitudes in US medical students. *Am J Clin Nutr* 84:655–62.
Thurston, V.C., et al. (2007). The current status of medical genetics instruction in US and Canadian Medical Schools. *Acad Med* 82:441–45.
Truswell, A.S. (2000). Family physicians and patients: is effective nutrition interaction possible?. *Am J Clin Nutr* 71:6–12.
University of British Columbia (2007) Nutrient metabolism and implications for health. Accessed 24 April, 2008, <http://www.landfood.ubc.ca/courses/outlines/fnh451.htm>.
University of Manitoba (2007). Faculty of Medicine, curriculum guide. Accessed 22 April, 2008, http://www.umanitoba.ca/faculties/medicine/media/Curriculum_Guide_for_upload_to_web.pdf.
US Department of Health and Human Services (2000). Healthy people 2010: understanding and improving health, 2nd edn. US Government Printing Office, Washington, DC. Accessed 1 January, 2008, <http://www.healthypeople.gov/Document/pdf/uih/2010uih.pdf>.
US Secretary's Advisory Committee on Genetics, Health, and Society (2007). Draft report on the US system of oversight of genetic testing. Accessed 4 December, 2007, <http://www4.od.nih.gov/oba/SACGHS/public_comments.htm>.
van Weel, C. (2003). Dietary advice in family medicine. *Am J Clin Nutr* 77(suppl):1008S–10.
Verhoef, M.J., et al. (2006). The scope of naturopathic medicine in Canada: an emerging profession. *Soc Sci Med* 63:409–17.
Vickery, C.E. and Cotugna, N. (2005). Incorporating human genetics into dietetics curricula remains a challenge. *J Am Diet Assoc* 105:583–88.
Wang, C., et al. (2004). Assessment of genetic testing and related counseling services: current research and future directions. *Soc Sci Med* 58:1427–42.
Watson, E.K., et al. (1999). The 'new genetics' and primary care: Gps' views on their role and their educational needs. *Fam Pract* 16:420–25.
Zdanowicz, M.M., et al. (2006). Pharmacogenomics in the professional pharmacy curriculum: content, presentation and importance. *Internat J Pharm Educ* 3:1–12.

Further reading

The Manitoba Pharmaceutical Association Self testing products guideline. Accessed 5 January, 2008, <http://www.napra.org/pdfs/provinces/mb/SELFTEST.pdf>.

CHAPTER

9

Advancing Knowledge Translation in Nutritional Genomics by Addressing Knowledge, Skills and Confidence Gaps of Registered Dietitians

Ellen Vogel, Ruth DeBusk and Milly Ryan-Harshman

OUTLINE

Nutrition and Genomics
ISBN: 978-0-12-374125-7

SUMMARY

Registered dietitians, as trusted and credible sources of food science and nutrition information, have a pivotal role to play regarding knowledge translation in nutritional genomics. There is a need to elucidate common gaps in the current capacities of dietitians in nutritional genomics to identify key success factors and barriers to moving forward and to propose strategies for advancing dietetics practice in nutritional genomics at national/international levels. A recent Canadian study on nutritional genomics examined opportunities and current gaps in dietetics practice, education and research. Data were collected through semi-structured key informant interviews and focus group interviews held with dietitians working in diverse practice settings across the country. Five themes emerged through data analysis: knowledge, skills and confidence gaps of registered dietitians; clinical validity and utility of predictive tests; population health versus medical perspectives; 'medicalizing' food and nutrition; and interdisciplinary and cross-sectoral collaboration. The knowledge translation role of the registered dietitian requires breadth and depth in various aspects of food and nutrition and the underlying genetic, biochemical and functional underpinnings. Registered dietitians must build on their foundational knowledge of food and nutrition science while increasing their understandings of human genetics including the ethical, legal and social issues associated with genetic testing.

INTRODUCTION

Advances in genomics are ushering in an era of personalized health, including personalized nutrition. Nutritional genomics is the foundation for personalized nutrition in which environmental factors, such as dietary and lifestyle choices, can be specifically tailored to meet an individual's needs with the goal of promoting optimal health, minimizing disease and, ultimately, maximizing genetic potential. The ability to reduce the impact of chronic diseases promises to decrease the economic burden of these disorders, while improving quality of life.

Not surprisingly, the initial applications of nutritional genomics will be health-oriented given that genes are central to the health and disease status of individuals and, thereby, populations. Fully integrating genomics into health care is expected to contribute to a fundamental shift towards increased efficacy in the management and prevention of chronic diseases. One potential outcome associated with this changing paradigm is the early detection of disease susceptibilities and the application of appropriate gene-directed preventive therapies. Many of the genes that predispose to chronic disorders can be positively influenced by dietary and lifestyle factors, making nutritional genomics key to disease prevention and health promotion (Box 9.1).

To realize the full potential of nutritional genomics in health care, consumers must understand their health issues and make informed decisions for themselves and their families (DeBusk and Joffe, 2006). In this chapter, it is posited that registered dietitians, as trusted and credible sources of food science and nutrition information, have a pivotal role to play regarding knowledge translation in the emerging era of diet–gene interactions. To fully embrace this new role, registered dietitians must build on their foundational knowledge of food and nutrition science while increasing their understandings of human genetics including the ethical, legal and social issues associated with genetic testing. The very important point to emphasize is that registered dietitians will need successfully to integrate these two disciplines into the science of nutritional genomics.

We foresee that registered dietitians will work in close collaboration with other nutritional genomic practitioners, including clinical genetic specialists and, in some countries, naturopathic physicians. Clinical specialists such as genetic physicians and genetic counselors are, however, currently limited in numbers and their expertise is largely focused on chromosomal abnormalities and highly penetrant genes (i.e., those responsible for cystic fibrosis, inborn errors of metabolism and sickle cell disease). Presently, nutrition science is rarely included in the education of clinical genetic specialists and, when incorporated into the curriculum, the coverage typically lacks the breadth and depth necessary to apply nutritional genomics in the health care setting.

It is proposed that registered dietitians who benefit from timely and strategic capacity building in nutritional genomics would be well positioned to address some of the current gaps, particularly in the area of knowledge translation (Box 9.2). The chapter begins by providing readers with pertinent background information on the education and training of dietitians in the USA, the UK and Canada. Next, it examines the evidence-base to ascertain how dietitians in these countries are moving to integrate nutritional genomics into practice. In this section, an overview of dietetic research conducted in the UK and the USA is initially provided, followed by a more detailed discussion of a Canadian study designed to identify opportunities and current capacity gaps – knowledge, skills, resources,

BOX 9.1

ASSOCIATIONS AMONG GENES, DIET, LIFESTYLE AND CHRONIC DISEASE

Nutritional genomics is based on key associations among genes and environmental factors, such as dietary and lifestyle choices. The interaction may be *nutrigenetic* where the focus is on the individual's genetic makeup and the ability to digest, absorb and use nutrients and other bioactive components in food for nourishment. Particular variations in a gene's sequence can alter the effectiveness with which the gene's protein product carries out these processes which, in turn, can affect the amount of a nutrient needed by that individual. Alternatively, the expression of many genes can be influenced by bioactive factors in the environment, which can vary from nutrients and other bioactive food components to estrogen and other steroids or environmental chemicals, such as polyaromatic hydrocarbons from char-grilled meats. This type of interaction is *nutrigenomic*. Changes in a gene's nucleotide sequence can alter the gene's response to particular bioactives.

Although each individual has the set of genes characteristic of the human species, the particular variations in gene sequence will differ between individuals. Thus, each person will have somewhat different nutrient requirements and respond somewhat differently to food components and other environmental factors, all of which can affect their susceptibility to developing a disease state. Nutritional genomics concerns assessing an individual's genetic variations and using this information, coupled to the gene variant–diet and lifestyle–disease associations, to develop therapeutic interventions that will improve disease management and provide effective approaches for disease prevention. Among the diet-related disorders that have been associated with genetic variations are many of the chronic diseases: cardiovascular disease, type 2 diabetes, cancer, osteoporosis, obesity and a wide variety of inflammatory disorders. A number of gene–diet associations have already been identified and numerous additional ones will emerge as the research moves forward. As this knowledge is integrated into clinical applications, nutritional genomics is expected to improve significantly the management and prevention of such diet-related diseases.

motivations – of registered dietitians in applying nutritional genomics to practice, education and research.

The objectives in writing this chapter are threefold:

1. to elucidate common gaps in the current capacities of registered dietitians in nutritional genomics;
2. to identify key success factors and barriers to integrating nutritional genomics into practice and

BOX 9.2

DEFINING KNOWLEDGE TRANSLATION

According to the Canadian Institute of Health Research (CIHR), 'knowledge translation is the exchange, synthesis and ethically-sound application of knowledge within a complex system of relationships among researchers and users' (http://wwwcihr-irsc.gc.ca/). Potential users of nutritional genomics knowledge range from researchers in a vast array of disciplines, policy makers/influencers, health service providers and the general public and consumer groups (e.g., the voluntary sector, educators, the media and non-governmental organizations) and the private sector.

3. to propose strategies for advancing dietetics practice in nutritional genomics at national and international levels.

The content of this chapter has been organized in two parts: Part 1 specifically addresses the first two objectives; Part 2 focuses on the third objective and presents recommendations.

PART 1: EDUCATION AND TRAINING OF REGISTERED DIETITIANS

Registered dietitians in the UK, the USA and Canada all graduate from a university or college after completing an undergraduate program of dietetics. The undergraduate dietetics curriculum integrates the principles of nutrition science, food science, biochemistry, physiology, food systems management, behavioral science and social science to achieve and/or maintain the health of an individual or group. The appropriate national credentialing agency must approve or accredit the program. The registered dietitian must also complete an approved, structured practicum. In the USA, the pre-professional must pass the five parts of the American Dietetic Association's (ADA) registration examination and maintain competency through continuing education. The five content areas examined include food and nutrition; clinical and community nutrition; education and research; food and nutrition systems; and management. Recently, the ADA is placing increased emphasis on genetics (Gilbride, personal communication, 2008), complementary care and reimbursement for medical nutrition therapy. In both Canada and the USA, increasing numbers of dietetic professionals are pursuing additional credentials such as a master's or doctoral degree and/or specialty certifications in a wide range of areas (e.g., acute care and chronic care disease management, health promotion, and performance-oriented areas).

In Canada, the USA and the UK combined, there are nearly 90 000 registered dietitians (Commission on Dietetic Registration, available at http://www.cdrnet.org/certifications/rddtr/rdbystate.htm). The UK has over 6600 registered dietitians (Health Professions Council, available at http://www.hpc-uk.org/); Dietitians of Canada (DC) currently reports over 5000 members (http://www.dietitians.ca) and, with more than 67 000 members, the ADA is the largest organization of food and nutrition professionals in the USA (http://www.eatright.org).

Surveys of Registered Dietitians in the UK and the USA

Results of a 2007 survey by Whelan and colleagues (Whelan et al., 2008) in the UK were similar to those obtained by a team of researchers in the USA (Wallace et al., personal communication, 2007). In the UK study, a questionnaire was mailed to 600 registered dietitians with a response rate of 69%. Of the 390 respondents, the majority had limited knowledge of nutritional genomics and lacked confidence in applying nutritional genomics in their work. The mean knowledge score concerning genetics and diet–gene interactions was 41%. Those registered dietitians who were using nutritional genomics to the greatest extent, however, had the greatest knowledge scores and confidence. These data suggest that enhanced education and training is necessary to increase the capacities of registered dietitians in nutritional genomics.

In the USA, a survey administered by the Center of Excellence for Nutritional Genomics at the University of California, Davis, in conjunction with the ADA and San Francisco State University, queried registered dietitians as to their awareness of nutritional genomics and perceived training needs in this area (Wallace et al., personal communication, 2007). The online survey, consisting of 20 closed- and open-ended questions concerning nutritional genomics, was sent to 10 000 dietetics professionals randomly selected from the ADA's member database. Data analysis is in progress, but preliminary results indicate that the majority of the 2300 respondents had been exposed to the topic of nutritional genomics, primarily through workshops and journal articles. Less than 10% of respondents reported that their clients asked questions about nutritional genomics and very few registered dietitians (less than 5%) felt themselves to be well informed on this topic.

The findings of both the UK and USA surveys indicate that registered dietitians currently have limited knowledge of the science of nutritional genomics. Further, there are gaps in dietitians' understandings of the potential application and impact of nutritional genomics on the profession, including the advanced knowledge and skills that will be required. These findings are consistent with the results of the HuGEM survey of genetics education of health professionals, including dietitians Lapham et al. (2000) and of Rosen et al. (2006) in their survey of the continuing education needs of registered dietitians with respect to nutrigenomics. The European situation is complicated in that professional qualifications for the dietitian vary from country to country (Cuervo et al., 2007). Despite this, Bouwman and Astley (2006) documented similar findings pertaining to the capacities of European dietitians in the area of nutritional genomics.

Canadian Situation: An Initial Analysis

A Canadian study supported by the Canadian Foundation for Dietetic Research and the Centrum Foundation was conducted from 2005 to 2007. An interdisciplinary research team contributed expertise in the following areas: nutritional genomics, biotechnology, human genetics, functional foods, ethical issues associated with nutritional genomics tests and information, chronic disease prevention, public health policy, interdisciplinary approaches to medical education and qualitative research methods. The team sought to increase awareness of new roles for registered dietitians in the emerging era of diet–gene interactions. The exploratory study was the first in Canada to examine dietitians' capacities in nutritional genomics. Multiple data collection techniques acted as an internal validity/credibility check (triangulation) such that data obtained by one method could be checked against data obtained by another method (Miles and Huberman, 1999). An overview of data collection strategies follows.

Key informant interviews

Twelve key informant interviews were completed. Key informants included national and international researchers, educators, policy makers, industry leaders and executives affiliated with large non-governmental organizations known for their expertise in areas related to the study.

Focus group interviews with registered dietitians

A focus group interview was held in each of six geographic regions defined by Dietitians of Canada. Interviews were conducted in Vancouver, British Columbia; Calgary, Alberta; Saskatoon, Saskatchewan; Toronto, Ontario; Montreal, Québec; and Halifax, Nova Scotia. Eight to ten dietitians participated in each focus group interview, facilitated by a trained researcher. Purposeful sampling techniques were used to bring together registered dietitians from diverse practice settings (e.g. hospital, community, private practice, industry) with dietitian-educators, researchers and policy makers. Participants were selected to capture a range of prior knowledge and/or experience in nutritional genomics.

Problem-based case study on nutritional genomics

A case study, entitled 'Diet and Risk of Cardiovascular Disease', was designed to stimulate conversation in the focus group interviews on a wide range of practice issues. Information provided to participants in advance included a synopsis of current evidence related to genetic polymorphisms, blood lipid levels and dietary interventions.

Researchers performed content analysis of the data and apparent themes and patterns were identified in a process of open coding (Strauss and Corbin, 1998). A qualitative data analysis software program (NVivo6) was helpful in sorting the data.

Results and discussion

A conceptual framework, based on a population health approach (Public Health Agency of Canada, 2002), was used to analyze the research findings at multiple levels. The population health approach directs interventions aimed at improving health toward broad, systemic determinants, many of which lie outside the traditional health care system. Collaboration among multiple sectors is a key component of this approach.

Researchers posited findings within the individual determinants (e.g. biological and behavioral) and the collective determinants (e.g. social, cultural, physical, economic and political) of healthy eating. The framework for the study was partly informed by an overview and synthesis of the determinants of healthy eating in Canada (Raine, 2005).

Five salient themes emerged through data analysis:

1. knowledge, skills and confidence gaps of dietetics professionals
2. clinical validity and utility of predictive tests
3. population health versus medical perspectives
4. 'medicalizing' food and nutrition
5. interdisciplinary and cross-sectoral collaboration.

A synopsis of findings regarding each thematic area follows.

Knowledge, skills and confidence gaps: 'What's a polymorphism?' Registered dietitians influence the development and promotion of consumer products, manage quantity food services in health institutions and provide information and counsel that allows decision-makers, including the consumer, to make informed judgments about food choices (Dietitians of Canada, 2007). According to information posted on the DC website (http://www.dietitians.ca/), registered dietitians are the most trusted source of information on food and nutrition in Canada.

To provide nutritional genomics advice, registered dietitians need to tailor dietary advice according to differences in genotype. According to Burton (2003a), successful implementation requires multiple competencies, including:

> an understanding of genetic risk, gene–environment interactions, and the possibilities and limitations of genetic testing as well as an awareness of some of the ethical, legal and social dimensions involved and an ability to communicate all of this to the patient. Further, professional [dietetics] expertise will be increasingly important as genetic tests and associated nutritional and other lifestyle advice become available directly to the consumer over the Internet or on the high street (Burton, 2003).

Most health professionals are not adequately prepared to integrate genetics into clinical practice (Collins, 1997; Scanlon and Fibison, 1995; DeBusk, 2002; Guttmacher et al., 2007; Aspinall and Hamermesh, 2007). Dietitians are *not* the exception. In a US survey of dietetics programs, only four of 82 directors reported that a genetics course was required (Vickery and Cotugna, 2005).

In the Canadian study, findings provided strong evidence of critical knowledge and skill gaps, which researchers partly attributed to the current lack of training in both human genetics and nutritional genomics in most undergraduate programs and/or dietetic internship programs. A recent graduate of a Bachelor of Science program explained in a focus group interview:

> I do not know what a polymorphism is. I know what a gene is... even though I've done a four-year degree and a one-year internship the only thing I remember is a little bit in first year anatomy and physiology.

Other focus group participants explained that current knowledge deficits in human genetics and gene–environment interactions made it '*almost impossible*' to appraise critically the scientific literature.

The following quotation is from a 'seasoned' practitioner with specialized training in nutritional genomics acquired primarily through postgraduate education. The dietitian referred to a shortage of continuing education programs in topics relevant to nutritional genomics. She went on to address challenges faced by clinicians seeking to enhance their knowledge base:

> I have the basics, but it's by working in the field, and going to many conferences in the [United] States, because there's nothing in Canada, nothing at all! And it's by reading, and catching up on this, and talking with the physicians, going to rounds, but we are far behind to practice at that level.

A second experienced dietitian also acknowledged confidence issues and a feeling of not being adequately prepared:

> I took two genetics courses, plus I took a graduate genetics course and I still find some of this difficult. I don't feel that I'm equipped ....

In this study, key informants and focus group participants agreed that it was important for the dietetics profession to explore this emerging area of practice, education and research. Further, the vast majority of dietitians interviewed recommended that the profession begin to develop competencies in this area. In the words of one dietitian:

> It [nutritional genomics] is a very far and wide-reaching area and I think professional dietetics really needs to jump on to it because, if we don't, it's certainly going to take off without us. I think that we need to be looking at all aspects from academia right through to in-patient and health care.

Some focus group participants predicted a shift to personalized nutrition would strengthen clinical outcomes, while increasing the profile of the dietitian:

> We won't be seen as the people with the diet sheet. . . . It will augment our level of rank and expertise.

In this vein, another clinician suggested:

> This [nutrigenomic tests] will increase access to objective data. Compliance often is improved when you can show those numbers and those markers to the patient and not just say, 'It's a bit high, it's a bit low'. The more information we have. . . . the more markers we can have . . . we can define very specific interventions that lead to improvement.

Dietitians currently providing outpatient counseling in group settings predicted critical resource shortages in health care systems already experiencing financial pressures. One focus group participant stated: 'More individualized counseling would definitely become a staffing issue'.

She elaborated:

> . . . 95% of my work is in the area of cardiovascular health, in a group setting, and I see that that wouldn't be feasible any longer. The waiting period would be *very* long and growing exponentially. I see the workload issues as being huge!

In focus group interviews, dietitians raised numerous practical issues that would need to be addressed before moving forward in nutritional genomics. For example, a clinician questioned how genetic test information would be translated into a realistic treatment plan:

> So now Jane has to go home and not only deal with cooking for her new genotype and [laughs] but she also cooks for her husband who has a completely different genotype and completely different dietary needs. . . . For me, a lot of it comes down to what recipes am I going to be giving Jane so that she can carry through?

In customizing meal plans for individuals based partly on their genotype, registered dietitians will face many knowledge translation challenges requiring practical, easily understood (i.e. plain language) and evidence-based approaches to dietary counseling.

Clinical validity and utility of predictive tests: 'Unregulated "gobbledygook" genetic testing results a concern of registered dietitians'

International media attention regarding the US General Accountability Office (GAO) investigative report on personalized genetic tests, testing companies, resulting nutritional advice and costs of tests and recommended supplements (Kutz, 2006) alerted both professionals and consumers to a wide range of problems. The investigation concluded that the tests that were examined misled consumers by offering inaccurate and/or highly generalized information. Further, several companies recommended that consumers purchase dietary supplements at inflated prices.

Dietitians need to understand the role of such tests in the marketplace and their nutritional implications in order to provide guidance to clients and/or consumers. A focus group participant shared her experience with non-scientific, genetic testing services currently available over the Internet. She expressed concerns about a lack of professional counseling, laboratory standards and the potential of false promises. The registered dietitian advocated increased government regulation to differentiate between valuable genetic testing services and results she described as genetic 'gobbledygook'.

Key informants spoke at length about the issues associated with (nutri)genetic testing. In most cases, they agreed that the 'evidence was still out'. In the words of one industry leader:

> We've become a little SNP [single nucleotide polymorphism]-centric on the value proposition of doing genetic tests and offering advice to people. ... I think we're actually a long way from that.

A second key informant emphasized that, in the area of (nutri)genetic testing, it was important for industry officials not to promise more than can be delivered:

> I want to be cautious about making this too readily available until we've moved down the road and some of these exotic tests and products we're dealing with become mainstream.

The comments of a third key informant touched on the potential public health applications of screening for genetic susceptibility:

> In order for a SNP, and a test for that SNP to be important, the SNP has to have the frequency in the general population that is really very, very high in order for it to be meaningful in a public health perspective. If you have one SNP it's easy to offer dietary guidance. If you have 2 SNPs you *might* be able to offer some dietary guidance, and if you have three SNPs, I suspect that you can no longer do it.

Reflecting on the conclusions of the GAO investigation report on personalized genetic tests, a fourth key informant suggested that it was important:

> not to let this sort of fledging initiative in nutritional genomics become ensnared in ethical debates and issues based on genetic testing, because it is such a small subset of what is potentially a very valuable morph of the nutrition science community into molecular nutrition.

Mathers (2004) cautioned that science cannot address whether one's genotype knowledge will encourage behavior change towards better health or create a sense of fatalism that leads to the adoption of high-risk behaviors. The notion of 'fatalism' in relationship to genetic testing arose in the focus group discussion. In the following statement a dietitian imagined how a client might respond to a personalized diet and lifestyle prescription:

> I can't do anything about it, nothing's going to work. . . . I'm just going to go along on my merry way and make no changes at all because I'm going to die, you know. Something's going to kill me, we get that all the time now but we're just starting to get the 'it's in my genes' message, because that is out there, the message is already out there.

Registered dietitians are well positioned to combat notions of fatalism by increasing awareness that nutritional genomics is concerned with identifying genetic variations that do not, in themselves, cause disease. Importantly, dietitians can utilize their highly tuned communication skills to convey the message that dietary and lifestyle changes may modify disease trajectories (DeBusk and Joffe, 2006).

Population Health versus Medicine Perspectives: 'Life choice or life chance?' Focus group discussion increased participants' and researchers' understandings of the political context surrounding nutritional genomics. For example, conversations highlighted the competing values, interests and beliefs of practitioners regarding the relationship between food and health. Lawrence and Germov (2004, p. 122) cite the work of Blane et al., (1996); Marmot & Wilkinson (1999) stating that, individuals who embrace a population health approach, 'generally subscribe to the view that the most powerful determinants of health are the social, economic, and

TABLE 9.1 Selected characteristics of population health and medical perspectives

	Population health perspectives	Medical perspectives
Scope/target	Population	Individuals
Cause of disease	Social and/or environmental conditions	Individual biology and behavior
View of health	Resource for living	Absence of disease (individual responsibility)
View of food	Prerequisite for health	Commodity to prevent or treat disease

Adapted from Selected characteristics of health promotion and medical paradigms (Lawrence and Germov, 2004).

cultural circumstances in which people live'. For these individuals, the priority interventions are those that promote food security.

In contrast, individuals who subscribe to a medical perspective typically believe population health can best be promoted and/or protected by preventing and treating diseases at the individual level. Lawrence and Germov (2004, p. 123) suggest that, according to the medical perspective:

> ... interventions aim to address the risk factors and genetic factors associated with disease by focusing on changing dietary intake. In this context, food is regarded as a commodity that may be modified to assist the dietary reform process.

Both perspectives (i.e. population health and medical) are summarized in Table 9.1, in relationship to the emerging science of nutritional genomics.

The following quotation from a focus group interview participant with an established track record in public health nutrition is congruent with population health perspectives where consideration of the social determinants of health (e.g. income, education, access to services) is paramount. The dietitian expressed concern that moving forward in nutritional genomics could place vulnerable individuals and/or populations in the position of 'choosing or losing' health:

> It sets up another gradient, right? Those who have resources, and it's usually the higher educated people who have a half decent income that allows them to purchase services, or healthy foods, and even things like specialty formula or the specialty vitamins. If the gap is getting wider, it could eliminate another [group], you know, middle-income, depending on how much these products would cost if they hit the market.

A second dietitian admitted to feeling confused and 'conflicted' as she questioned the impact of nutritional genomics on the social determinants of health. She observes that lifestyle choices are often significantly influenced by life circumstances:

> I want to see people eat well and choose healthy foods. But as a public health worker I know that the things that will make a difference will be social assistance rates, minimum wage, assistive education. .. All those things would make people far healthier. I don't see nutritional genomics helping the social determinants of health.

In a chapter examining the politics of functional foods and health claims, Lawrence and Germov (2004, p. 124) cautioned against, 'extrapolating scientific evidence' from the population health perspective to the medical perspective. In their analysis of the political constructs underlying functional foods and health claims, the authors conclude: 'It is more appropriate to look at the expectations. . . . within the context in which they were developed: in relation to the potential effects on individuals' p. 124. Perhaps this sage advice pertains to the emerging science of nutritional genomics as well?

'Medicalizing' food and nutrition: 'Food is more than just nutrients'

Specialized foods are not a new idea – the marketplace already offers low-fat foods, diabetic foods and salt-reduced foods. Experts predict the development of new foods designed to target specific physiological processes or genetic limitations. A rapid increase in personalized foods may be facilitated by identification of biomarkers that improve individual assessment and the emergence of technology platforms such as mass spectrometry, nuclear magnetic resonance and high-resolution chromatography that can easily and accurately measure many metabolites. Also important will be the ability to link metabolites to phenotypes and to link food composition to metabolites (German et al., 2004).

Concerns have been expressed regarding 'medicalizing' the food supply (i.e. producing foods that approximate drugs) and the inherent risks in minimizing the socio-environmental influences affecting health. According to Lawrence and Germov (2004):

> The medicalization of food involves treating food like a drug with therapeutic properties that are able to prevent diseases. Such a view represents a pathologized and reductionist approach to health promotion and food consumption. The likely outcome is that the individual will be blamed for any diet-related illness [victim-blaming] since the mode of prevention simply becomes a matter of food consumption choices p. 124.

Some focus group comments suggested that dietitians shared some of the concerns articulated above by Lawrence and Germov (2004):

> You have to think of food as how we relate to people, it affects our culture, it affects our ethnicity, it affects. .. how we inter-relate socially, and the thought of that. .. being broken down into alleles and nutrients. . . . I'm just trying to get my head wrapped

around this. ... it's emphasizing one aspect of nutrition, food as a vehicle of getting our nutrients, and losing all these other aspects of food.

Another dietitian cautioned that producing foods that approximate drugs and/or overemphasizing the health benefits of foods, could potentially interfere with the social dimensions of preparing food and eating together:

So, just because the science is there doesn't mean that that's where people are going to be. And as we always say when we have our beer and pizza on Friday nights, it's not about physiological needs at all. I think it [medicalizing food and nutrition] takes away the joy of food and that's going to be a tough, tough sell! To most people that would not be an equitable balance between you're going to tell me to eat these certain foods to annihilate my risk of genetic disease but you're going to take away the foods I really like ...

Interdisciplinary and Cross-Sectoral Collaboration: 'Finding new ways of working' When contemplating a roadmap for nutritional genomics over the next decade, an experienced clinician emphasized the importance of building new interdisciplinary bridges with experts in genetics, food science, cultural anthropology, molecular biology, pharmacy and public health:

I think it will be important that there are strong, multidisciplinary teams. ... I know how to translate some of the information but I always like to work with the geneticist because there are some borderline decisions that you need to make. ... it's important that you have people who know their stuff, people you can work strongly with because I might have a solution, you might have a solution, but at least you know you can remain strong and come to a decision to the best of your interpretation. And this is always your science evidence background.

Key informants echoed the importance of strong interdisciplinary teamwork. One individual stated:

Dietitians will provide the personal nutrition advice based on their knowledge of nutritional genomics. However, we'll need a collaborative approach among different health care providers, physicians, pharmacists, and all other nutritional health professionals.

Both key informants and focus group participants suggested that media involvement was a key success factor together with consumer involvement. One dietitian spoke of the need for a broad social marketing campaign to increase public awareness of nutritional genomics:

For me, it comes back to how to involve the media? I'd like to see a broad social marketing campaign, particularly as we move into this era of 'different strokes for different folks'. We're now going to have situations where there are a million answers, there's an answer for every person. . . . and people are going to need to learn differently.

Finally, a dietitian leader, with expertise in food and nutrition policy at the national level, concluded the key informant interview by suggesting that *'we learn from our past mistakes'*. In the interview, she specifically referred to ongoing challenges in Canada associated with reaching stakeholder convergence on biotechnology. She emphasized that intersectoral collaboration and new business models for working with industry partners, were both required to address some of the barriers to moving forward.

PART 2: STRATEGIES FOR ADDRESSING CAPACITY GAPS

In this section are outline strategies for advancing dietetics practice in nutritional genomics with special emphasis on knowledge translation. It begins with envisioning the potential of the registered dietitian as a nutritional genomics practitioner. Next, the focus is on the role of education and training in building dietitians' capacities to move forward. The section then reviews the current gaps and barriers facing registered dietitians nationally and internationally and concludes with recommendations for action.

The Role of the Registered Dietitian in the Genomics Era

The potential applications of nutritional genomics to both food science and nutritional science and practice are wide ranging and include basic and applied research, clinical and public health applications, education and policy development. Presently, research is focused on identifying gene–diet/lifestyle associations and delineating the details of these interactions. It is anticipated that the focus will shift to clinical research that investigates the efficacy of diet and lifestyle interventions on the progression or prevention of particular chronic disorders. Registered dietitians with expertise in nutritional genomics will play key roles on research teams conducting studies, including clinical trials, and in the development of nutrigenetic tests.

Dietetic professionals with food science expertise will be engaged in identifying and isolating bioactive food components that influence gene expression. These bioactives may be added to commonly eaten foods and/or encapsulated into dietary supplements. Further, it is anticipated that new crops, enriched with specific bioactives, will soon be developed. As the basic research foundation expands, the practical applications in both nutrition and food science will expand.

A major application will be the use of genetic technology to detect variations within an individual's genetic makeup that have been associated with increased susceptibility for developing a chronic disorder. Registered dietitians will need to use foundation knowledge regarding the mechanisms of disease, together with the underlying genetic bases, to develop appropriately targeted, effective therapies. This will require new and/or enhanced competencies in the following areas:

1. incorporating nutrigenetic testing into nutritional assessment
2. explaining the results and their implications to patients/clients and their families
3. translating the clinical findings into food, supplement and lifestyle choices
4. assisting clients in making the needed behavior changes through coaching, food selection and food preparation.

Clearly, the potential knowledge translation role of the registered dietitian in the coming era of genomics will be a demanding one, requiring breadth and depth in various aspects of food and nutrition and the underlying genetic, biochemical and functional underpinnings. In time, nutritional genomics is expected to be integrated into much of the dietitian's activities. The first visible evolution of the dietitian's role will likely be in clinical nutrition – a registered dietitian with advanced practice expertise in nutritional genomics analogous to the advanced registered nurse practitioner or physician assistant. These postgraduate professionals enjoy increased autonomy and respect commensurate with their advanced credentials and the enhanced value of their contribution to health care. Skipper (2004) and Skipper and Lewis (2005, 2006a, 2006b) explored many of these issues and proposed this type of advanced practice model for the registered dietitian.

It is envisioned that the registered dietitian nutritional genomics practitioner will function as the 'go-to' member of the patient/client's team of health care advisors. This individual will add value by providing guidance on all aspects of nutritional genomics, from nutrigenetic testing to the development of gene-based therapeutic and preventive interventions. The advanced-practice registered dietitian will integrate various types of genetic information into the nutrition assessment to generate a comprehensive picture of the individual's potential risk for disease development. By developing gene-directed strategies for modulating diet and lifestyle factors that interact with gene variants, it will be possible to increase the efficacy of nutritional interventions. Further, such a practitioner is well positioned to play a catalytic role in shifting health care from its current emphasis on disease management to disease prevention and health promotion.

Such a dietetics practitioner does not currently exist but the need for an advanced practice nutritional genomics dietitian is increasingly recognized. Ongoing discussions among practitioners, professional dietetics organizations and universities are seeking to define the need and the best career path for moving forward. The scientific education and skills training required suggest that the program will need to include an advanced degree focused on clinical practice, with practical experience in nutritional genomics as an essential component of the degree. An advanced degree alone is not sufficient for establishing the registered dietitian as an expert in nutritional genomics. There will also need to be a recognized credential for dietetic professionals with clearly articulated competencies and rigorous criteria applied to ensure the competencies have been achieved.

The Role of Education in the Development of a Nutritional Genomics Practitioner

Education is a common thread related to all of the action steps described above. In addition to a solid foundation in the sciences underlying food and nutrition and its applications, the registered dietitian will need to develop expertise in genetics and in the role of genes in health and disease. In-depth knowledge of the implications of genetic variation on the ability to use nutrients to promote health, both to prevent and to manage disease will be needed, together with skills to develop and implement therapeutic approaches that appropriately match diet and lifestyle to the client's genetic makeup. Importantly, the registered dietitian will need to enhance existing communication skills and develop the leadership attributes necessary to establish this emerging discipline within the health care system.

Dietetic students, and the professors who teach them, will need specialty training in nutritional genomics, beginning with the fundamentals and continuing through to the advanced graduate level. Simultaneously, accessible and needs-based continuing education programs must be developed for 'seasoned' practitioners (Rosen et al., 2006). Successful implementation of continuing education programs is challenging and must consider career and family demands on the practitioner. It is anticipated that distance learning modalities will play a major role in the continuing education of registered dietitians.

Barriers and Opportunities to Moving Forward

Building on the foundation of the registered dietitian credential is an expeditious route to developing nutritional genomics practitioners with the needed versatility. Among the gaps that must be addressed for registered dietitians to realize this potential are:

- education in genetics/genomics in general, in the integration of nutrition and genetics into nutritional genomics in particular, and in

translating complex nutritional genomic science into practical applications for consumers
- hands-on training in the practical applications of nutritional genomics, with emphasis on the clinical aspects
- the development of a profitable business model for a registered dietitian nutritional genomics advanced practice
- basic business-related skills that will enable registered dietitian nutritional genomics practitioners to market themselves and their practices in a financially sound manner
- in countries with government-sponsored health care plans, the integration of nutritional genomics into the present framework of such plans.

The education and training components will be particularly challenging gaps to address initially due to the lack of registered dietitian nutritional genomics practitioners available for teaching and practice mentoring. Practitioners must have hands-on experience with nutritional assessment that includes nutrigenetic testing. The assessment must be translated into effective diet and lifestyle interventions and communicated to the client in ways that promote sustained behavior change and enrich enjoyment of food.

At present, there are few innovative service delivery models involving a registered dietitian nutritional genomics practitioner. In the USA, where health professionals are increasingly independent practitioners, neither nutrition services nor genetic services are readily reimbursed by insurance companies or other third-party payers. There are, however, isolated 'promising practices' that may serve as useful models (Box 9.3).

Under the publicly insured health plans of countries such as Canada and the UK, the issues associated with incorporating nutritional genomics services into the health care continuum will differ from the challenges faced in the USA. The need to demonstrate cost-effectiveness of these services will be common to practitioners and policy makers in all countries. The registered dietitian nutritional genomics practitioner will be involved 'in making the case' for including these specialized services in the health care offered to citizens. Advocacy, particularly at a national level, is not a role that most dietitians have traditionally filled. Thus, they will likely require training and skill development in advocacy to be successful at the policy table.

Potential barriers to addressing these identified gaps and realizing the full potential of registered dietitians as nutritional genomics practitioners include:

- a lack of trained educators who can teach nutritional genomics to dietetic students and practicing dietitians
- the need for innovative approaches to learning so that educators and practicing dietitians can access the requisite education and training

BOX 9.3

SPOTLIGHT ON THE REGISTERED DIETITIAN IN AN INNOVATIVE NUTRITIONAL GENOMICS PRACTICE MODEL

As the prevalence of chronic disease rises, the complexity of chronic care and the attendant increased demands on health care systems are causing many to rethink traditional models of care. One innovative model that is developing harnesses the valuable contributions that different types of health care professionals can make to patient care. Teams of health care professionals working together are enabling each to contribute what each does best while affording the patient the collective wisdom of more than one practitioner and potentially decreasing the burden on the primary care physician. The registered dietitian (RD) nutritional genomics practitioner is expected to play an important role in this evolving model.

The chronic care model is based in the discipline of functional medicine, a sub-discipline of conventional medicine with an emphasis on identifying the root cause of a patient's dysfunction and targeting therapy to the underlying genetic and biochemical mechanisms leading to the dysfunction (See www.functionalmedicine.org for more information). Functional medicine focuses on the genetic and biochemical individuality of each patient and the role of food and other environmental factors in the health of the individual. In this model, the RD nutritional genomics practitioner is viewed as an essential element of the chronic care team. This practitioner will be responsible for performing a comprehensive clinical nutrition assessment, which will include a physical assessment, diet-and-lifestyle assessment, extensive family history, nutritionally-relevant laboratory work and nutrigenetic testing, where appropriate. The extensive nutrition assessment and accompanying testing enables the physician to move quickly to the physical exam, follow-on laboratory testing and diagnosis. Other practitioners may also see the patient and contribute recommendations for the patient's care. The team meets to discuss each patient and together arrives at a recommended plan of care. This model is now being tested and eagerly awaited by those who recognize the value of a team effort in caring for those with one, or often multiple, chronic disorders.

The RD nutritional genomics practitioner is able to step into this new position by virtue of the strong foundation of food science and nutrition science, coupled to the breadth of training that ranges from research to clinical nutrition to food selection and preparation. RDs in this setting will regularly draw on their background in biochemistry and metabolism and incorporate both into clinical applications. By further acquiring knowledge in genetics, molecular nutrition and nutritional genomics and integrating these into their extensive foundation in the food- and nutrition-related sciences, RD nutritional genomics practitioners will be well positioned to meet the growing demands of the coming health care era.

- a lack of knowledge in the marketplace, among health professionals and consumers, about the role of genes in health and disease and the value of accessing expertise in nutritional genomics
- existing health care systems that rely upon third-party reimbursement for services or are structured as government-sponsored plans
- legal barriers stemming from limited scope of practice in most venues
- variability among jurisdictions concerning dietitians' professional standing. It is anticipated that the elevation of the registered dietitian to an advanced nutritional genomics practitioner will proceed more smoothly in venues where dietitians are held in high regard.

The capacity gaps identified in Part 1 of this chapter share many commonalities, with the possible exception of insurance reimbursement issues in the USA. However, publicly insured health systems face similar challenges with respect to including nutritional genomics services in routine nutritional assessment as a tool for disease management and health promotion.

Common Problems, Common Solutions

None of the barriers identified in this chapter is insurmountable and may be addressed with concerted effort by stakeholders. At a minimum, consideration should be given to including genetics as a core science in the undergraduate dietetics curriculum and to incorporating genomics in general, and nutritional genomics in particular, into the biochemistry, physiology and metabolism courses at both the undergraduate and graduate levels. Coursework pertaining to diet and disease must address the root cause of disease – the underlying genetic and metabolic bases for function and dysfunction – so that intervention therapies are highly targeted to the specific parameters leading to dysfunction and disease.

The dietetics profession, through its leader organizations nationally and internationally, can facilitate this transition by emphasizing critical thinking, mandating the inclusion of genetics and nutritional genomics at the undergraduate and graduate levels and requiring that these topics be included in the registration examinations that constitute part of the credentialing process in the USA.

Appropriately educated, trained and accredited, the registered dietitian nutritional genomics practitioner may have the added challenge of developing a profitable practice. In the USA, the nurse practitioner and the physician assistant may serve as useful service delivery models. Typically, these practitioners must work under the guidance of a licensed physician, but are free to work either directly with a physician or to establish their own practices. These professionals have key advantages not currently enjoyed by the registered dietitian, namely that their services are readily recognizable by the public and reimbursable through insurance. Thus, these practitioners are 'attractive' to physician employers, a fact that, in turn, facilitates the establishment of a financially sound practice for both the physician and the nurse practitioner or physician assistant.

In contrast, the business model for the registered dietitian nutritional genomics practitioner will likely need to be fee-for-service rather than based on insurance reimbursement. Establishing such a practice is more difficult, but quite feasible and potentially more financially lucrative in terms of fee levels and reduced staffing needs. Fee-for-service practitioners must, however, possess strong health care skills that translate into positive patient outcomes that are readily demonstrated. Practitioners must also be able to market themselves and their practices effectively. Given the lack of public and health professional familiarity with the value of nutrition therapy in general and the added value of incorporating nutritional genomics into nutrition therapy, practitioners will need to educate potential clients (patients and the health professionals who refer them) and to market themselves and their practices to establish a financially viable practice.

As the demand for nutritional genomics services expands, revisions to the scope of practice for registered dietitians will need to expand. One of the challenges for dietitians has been that the scope of practice is often limited to activities that do not, in general, take advantage of the extensive education, training and experience of the registered dietitian. By extending the scope of practice for the dietitian with advanced credentials as a nutritional genomics practitioner, the opportunity exists for broadening current professional practice guidelines. The potential benefits include attracting and retaining the most promising dietetic practitioners who are increasingly tempted by enhanced career satisfaction, autonomy, respect and financial remuneration afforded by other health professions, such as physicians. In an era that will increasingly rely on the capacities of practitioners to translate complex science and technology into lay terms, attracting and retaining 'the best' will serve the dietetics profession well.

Progress to Date and Suggested Actions

Progress is being made in a number of these areas, particularly with respect to professional education. Recently, genomics was included as one of four priority areas within the ADA's strategic plan (www.eatright.org website). ADA-related annual meetings at the national and state levels have included presentations on genomics and nutritional genomics for the last several years. The ADA is partnering with the National Coalition for Health Professional Education in Genetics (NCHPEG), the NHS National Genetics Education and Development Centre and the British

Dietetic Association to develop a web-based genetics module for dietitians that is expected to be available in 2008. In Canada, the Canadian Foundation for Dietetic Research (CFDR) was created in 1991 by Dietitians of Canada (DC) to support applied nutrition and dietetic research. In 2005, CFDR, in partnership with the Centrum Foundation, funded the work by Vogel and colleagues examining dietitians' capacities in the emerging science of nutritional genomics. Recently, DC's Practice-Based Evidence in Nutrition (PEN) initiative included a background document on nutritional genomics and dietetic practice as an initial step in the development of a knowledge pathway for members.

At the academic level, universities have begun to offer nutritional genomics education for dietitians. In the USA, the University of California, Davis has offered an online course in nutritional genomics for graduate students and postgraduate dietitians. The University of Medicine & Dentistry of New Jersey has an established online clinical nutrition graduate program and offers nutritional genomics as part of its curriculum. Discussions are underway involving academic institutions, non-profit organizations and for-profit businesses concerning the development of a training program for nutritional genomics practitioners to serve the early need for practitioners prior to the establishment of an academic degree program. Some of the private companies that are pioneering the development of nutrigenetic testing are also providing basic education in nutritional genomics for health professionals and consumers, as is the US government's National Institutes of Health. Food manufacturers are becoming increasingly interested in the applications of nutritional genomics to food and have expressed interest in developing information portals to assist with the education of consumers and health professionals.

Nutritional genomics expertise exists within the research community and, to a more limited extent, within the health care community. University dietetics programs and professional dietetics organizations can take advantage of the expertise that presently exists in several ways. When scientists and clinicians are available locally, many are willing to speak to classes and meet with students. Some may be willing to participate in courses, either as guest lecturers or adjunct professors. Distance learning technology, stand-alone 'webinars', continuing education short courses and full-term credit courses are ways to incorporate expertise not directly available to the university or professional association.

There will be opportunities for universities and associations to collaborate with businesses. Research collaboration has been a long-standing opportunity between scientists at different academic institutions and between academic and corporate laboratories. It is timely to begin developing partnerships with nutritional genomics scientists and practitioners who have knowledge and experience that would be valuable to dietetics students at all stages of development. The potential exists for internship opportunities for students, particularly within corporate or practice locations where students can be exposed to laboratories performing genetic testing and practices in which genetic testing is used. In situations where a practitioner uses traditional genetic testing, it would be helpful for the student to observe the process, including informed consent, data interpretation and counseling. Further, rotations through various clinics in which genetic analysis is incorporated into the client assessment would be useful. Rotations that involve exposure to genetic counselors are instructional and bring a heightened appreciation for the important contribution of these health professionals.

The development of academically-based education and training programs takes time and is not likely to be in place soon enough to meet the needs of the first wave of nutritional genomics applications. An entrepreneurial entity (e.g. business, non-profit organization, or academic entity such as a university extension service) could perhaps develop an interim certification and training program for nutritional genomics practitioners. If developed with input from knowledgeable individuals from the academic and business sectors, such a program could mature into a graduate degree program, including an advanced practice credentialing program for registered dietitian nutritional genomics practitioners.

Raising the awareness of consumers and health care professionals of the role and value of the nutritional genomics practitioner will be an important aspect of the acceptance of a new type of practitioner. Canadian, UK and US dietetic associations will face this challenge and collaborative efforts among these organizations would likely pay greater dividends than each association working independently. Clearly, there are numerous challenges ahead as nutritional genomics emerges as an essential discipline, however, challenges often bring with them opportunities. As the potential of nutritional genomics in improving health outcomes becomes evident, the registered dietitian with well-developed skills and a proven track record in working with diet–gene interactions, can play a central role in disease management and prevention.

CONCLUSION

Currently, the knowledge and skills of registered dietitians regarding diet–gene interactions remains limited, yet experts predict that nutritional genomics will impact virtually every sub-discipline of the profession. It is essential that the education and training of registered dietitians include in-depth knowledge of nutritional genomics and its potential dietetic applications so that practicing dietitians and pre-professionals will have the requisite knowledge, skills, resources and confidence to realize the potential of this dynamic field.

Initially, we envision a practical program that builds on registered dietitians' current depth and breadth of food science and nutrition science knowledge and skills. To this foundation, we propose adding coursework in human genetics and nutritional genomics in particular. Skill enhancement is necessary related to incorporating nutrigenetic testing and family history into nutritional assessment and translating this information into effective dietary and lifestyle approaches for the client. Over time, this program would be augmented with a formal degree program and subsequent certification that would generate the type of practitioner needed for this new era – a registered dietitian who enjoys the autonomy, professional respect and income commensurate with the valued contribution of such a practitioner to the health care team.

ACKNOWLEDGMENTS

The authors are grateful to expert reviewers in Canada, the USA and the UK for helpful comments on earlier drafts (Kim Raine, PhD, RD; Judith Gilbride, PhD, RD; Sharonda Wallace, PhD, RD; and Kevin Whelan, PhD, RD).

References

Aspinall, M.G. and Hamermesh, R.G. (2007). Realizing the promise of personalized medicine. *Harv Bus Rev* :109–17.

Blane, D., Brunner, E. & Wilkinson, R. (eds). (1996). Health and Social Organization: Towards a Health Policy for the Twenty-First Century. Routledge, New York.

Bouwman, L.I. and Astley, S. (2006). Exploring actors' perspectives about nutrigenomics based personalised nutrition. Project no. FP6-506360, priority five. The European Nutrigenomics Organisation.

Collins, F.S. (1997). Preparing health professionals for the genetic revolution. *J Am Med Assoc* 278:1285–6.

Cuervo, M., Brehme, U., Egli, I.M., et al. (2007). Nutrition, dietetics and food sciences degrees across Europe. *Ann Nutr Metab* 51:115–18.

DeBusk, R.M. (2002). *Genetics: the nutrition connection*. American Dietetic Association, Chicago.

DeBusk, R.M. and Joffe, Y. (2006). *It's not just your genes!*. BKDR, Inc., San Diego.

German, J.B., Yeretzian, C. and Watzke, H.J. (2004). Personalizing foods for health and preference. *Food Technol* 58:26–31.

Guttmacher, A.E., Porteous, M.E. and McInerney, J.D. (2007). Educating health-care professionals about genetics and genomics. *Nat Rev Genet* 8:151–57.

Kutz, G. (2006). *Nutrigenetic testing: tests purchased from four web sites mislead consumers*. US General Accountability Office, Washington, DC.

Lapham, E.V., Kozma, C., Weiss, J.O., Benkendorf, J.L. and Wilson, M.A. (2000). The gap between practice and genetics education of health professionals: HuGEM survey results. *Genet Med* 2:226–31.

Lawrence, M. and Germov, J. (2004). Future food: the politics of functional foods and health claims. In A sociology of food and nutrition: the social appetite, 2nd edn, pp. 119–47. Oxford, South Melbourne.

Marmot, M. & Wilkinson, R.G. (eds). (1999). Social Determinants of Health, Oxford University, Oxford.

Mathers, J.C. (2004). The biological revolution – towards a mechanistic understanding of the impact of diet on cancer risk. *Mutat Res* 551:43–9.

Miles, M.B. and Huberman, A.M. (1999). *Qualitative data analysis: an expanded sourcebook*, 2nd edn.. Sage, Thousand Oaks.

Public Health Agency of Canada (2002). What is the population health approach? Available from: http://www.phac-aspc.gc.ca/ph-sp/phdd/approach/index.html Accessed 15 November, 2007.

Raine, K. (2005). Determinants of healthy eating in Canada: an overview and synthesis. *Can J Public Hlth* 96:8S–15S.

Rosen, R., Earthman, C., Marquart, L. and Reicks, M. (2006). Continuing education needs of registered dietitians regarding nutritional genomics. *J Am Diet Assoc* 106:1242–5.

Scanlon, C. and Fibison, W. (1995). *Managing genetic information: implications for nursing practice.* American Nurses Publishing, Washington, DC.

Skipper, A. (2004). The history and development of advanced practice nursing: lessons for dietetics. *J Am Diet Assoc* 104:1007–12.

Skipper, A. and Lewis, N.M. (2005). A look at the educational preparation of the health-diagnosing and treating professions: do dietitians measure up?. *J Am Diet Assoc* 105:420–27.

Skipper, A. and Lewis, N.M. (2006a). Using initiative to achieve autonomy: a model for advanced practice in medical nutrition therapy. *J Am Diet Assoc* 106:1219–25.

Skipper, A. and Lewis, N.M. (2006b). Clinical registered dietitians, employers, and educators are interested in advanced practice education and professional doctorate degrees in clinical nutrition. *J Am Diet Assoc* 106:2062–6.

Strauss, A.L. and Corbin, J. (1998). *Basics of qualitative research: grounded theory procedures and techniques*. Sage, Thousand Oaks.

Vickery, C.E. and Cotugna, N. (2005). Incorporating human genetics into dietetics curricula remains a challenge. *J Am Diet Assoc* 105:583–88.

Whelan, K., McCarthy, S. and Pufulete, M. (2008). Genetics and diet-gene interactions: involvement, confidence and knowledge of dietitians. *Br J Nutr* 99:23–8.

Further reading

Burton, H. (2003a). *Dietitians education: workshop report*. The Wellcome Trust and the Cambridge Public Health Genetics Unit, London.

Burton, H. (2003b). Letter to the editor. *J Hum Nutr Dietet* 16:47–8.

Canadian Institute for Advanced Research (2002). Knowledge translation strategies for health research (Archived). Available at http://www.cihr-irsc.gc.ca/.

CHAPTER 10

Understanding Hopes and Concerns about Nutrigenomics: Canadian public opinion research involving health care professionals and the public

Alan Cassels

OUTLINE

Nutrition and Genomics
ISBN: 978-0-12-374125-7

SUMMARY

Canadians will readily admit they cannot knowledgeably assess nutrigenomic technology, its benefits or its potential harms because of the newness of the technology. Canadian researchers conducted focus group research to investigate reactions of members of the lay public and health care professionals to information about nutrigenomics science, nutrigenetic testing and associated services. Sixteen focus groups were conducted in cities across Canada: 12 focus groups involved consumers and four focus groups involved Canadian health care professionals (dieticians, physicians, naturopaths and pharmacists). Focus group participants were exposed to an Associated Press newspaper article, a mock nutrigenomics company website and an informational handout. Consumers were interested and curious about nutrigenomics, yet tended to grow increasingly skeptical towards the technology as the information exposures became more commercial in tone and substance. Unprompted, consumers and health professionals repeated a strong preference for government attention to nutrigenomics, both in communicating benefits and risks of nutrigenetic testing and related services, and in ensuring adequate standards of health protection. In communicating this new area of scientific knowledge to consumers, communication materials should be based on a solid understanding of: (1) how consumers and health professionals currently view the significance of (nutri)genetic tests to their lives and the lives of their families; (2) the sources of health advice consumers find most trustworthy; and (3) motivations for consumers to seek health advice, especially regarding nutrition and genetics. This chapter advances understanding of how to shape appropriate and comprehensive communications around nutrigenomics and stresses the importance of government and health professional cooperation in helping to fill a current information vacuum on the subject.

INTRODUCTION

As an emerging field, nutrigenomics faces challenges in communicating complex information to the public about the relationship between genetics, nutrition and health. This challenge is complicated by the rapidly evolving nature of nutrigenomic science, the growing public reliance on the Internet as a major source of health information and consumer preferences for access to some health services in the private marketplace, often independent of primary health care counseling. Despite a growing body of public opinion research studying citizen and consumer perceptions regarding genetics issues, there has yet to be a thorough assessment of knowledge and attitudes regarding nutrigenomics. Public perceptions of benefits and harms related to nutrigenetic testing are largely unknown and Canadian consumers, like those in many other countries, are, at present, largely unexposed to nutrigenomics.

While the science of nutrigenomics is relatively new, many of the underlying concepts and related issues have been the subject of discussion, controversy and public opinion research over the past decade. Key issues include the impacts of genetic makeup, environment exposures and lifestyle on health and disease, the desire and belief in personal control over health, the role of basic and enhanced nutrition in attempting to prevent diseases, the promotion of genetic screening tests, treatment of healthy people to avoid future chronic diseases and the benefits and harms of new technologies upon which this science relies (such as computing technology, genetic testing, nutritional manipulation of foods, gene patenting and development of biobanks).

This chapter explores issues of communicating with the public about the science of nutrigenomics and its benefits and harms. It discusses focus group research conducted in Canada involving members of the public and health care professionals. These focus groups aimed to address several overarching questions: among Canadians, what is the level of knowledge and awareness of nutrigenomics; how do people perceive the relevance of nutrigenomics to their lives or the lives of their families; when exposed to information about this new technology, what benefits or harms do they identify; and how do they think benefits can be maximized and risks minimized?

Understanding Canadians' awareness and attitudes about nutrigenomics helps to identify knowledge gaps that have important implications. Knowing what gaps exist in public knowledge about nutrigenomics can help to ensure that future development of the field is communicated in ways consumers and professionals find meaningful. Moreover, knowing where gaps in understanding lie can aid in the development of targeted, strategic investments to close knowledge gaps. Current investments in nutrigenomics are largely directed towards advancing the development of commercial initiatives while little is being done to consider the public health and public communications implications. One of the strongest and most consistent themes that emerged from these focus groups was that nutrigenomics testing should not be left exclusively to the marketplace. Respondents recommended that public health agencies in Canada should support health professionals in dealing with their patients' concerns regarding nutrigenomic testing or in helping serve patients in general, especially in terms of providing unbiased information on nutrigenomics for members of the general public, sponsoring research into the area of nutrigenomics through universities or research institutions with no vested interests.

COMMUNICATION CHALLENGES FOR NUTRIGENOMICS

With regard to public communication around nutrigenomic science and services, several key features pose challenges:

- lack of common understanding of concepts of 'health', 'environment', 'risk reduction' and 'disease'
- scant public opinion research and data, but what does exist shows a very low awareness of nutrigenomics among the general public and health professionals
- promotional material about nutrigenomics, especially that from some companies, makes misleading or exaggerated claims
- information that does not exaggerate the state of the science cannot yet point to any impressive achievements of nutrigenomics
- in some cases, nutrigenetic testing seems to be directed towards selling 'personalized' diet supplements, which may promote skepticism among some consumers.

Growing public interest in more holistic approaches to health, increasing use of complementary and alternative health products and therapies and regular exposure to genetic research findings and discussion through mass media create an environment that may foster consumer interest and openness to nutrigenomics.

The greatest challenge underlying communication of complex gene–environment interactions as it relates to accurate identification of the risk of chronic disease is the number of biases, fallacies and errors in the way that patients, the general public and health providers think about disease susceptibility. Susceptibility, or predisposition to disease, is typically framed in the language of risk and, more importantly, relative risk. Current models of communication about risk and the psychological processes that underlie risk perception raise the following issues that arise in the communication process (Klein and Stefanek, 2007):

- problems with use of numerical information (innumeracy)
- problems with the readability of materials (literacy)
- problems with cognitive processes (e.g. time saving heuristics)
- motivational factors (e.g. loss and regret aversion) and emotion.

Intermediaries, such as professionals and the media, have difficulty with 'risk' and 'predictive' value of tests and treatment for individuals and populations. Understanding the differences between relative and absolute risk, the differences between statistical significance and clinical significance, and between surrogate and clinical markers are all examples of issues where varying levels of understanding can seriously color the perception of risk.

PREVIOUS PUBLIC OPINION RESEARCH

Some recent research has examined Canadians' concerns about food and health. Decima Research (2006) studied the level of awareness and knowledge of Canadian consumers about the links between diseases and foods/food components, the most desired types and sources of information about foods that can contribute to a healthy lifestyle, attitudes towards functional foods and nutraceuticals, the determinants of these attitudes and identification of the products or product categories of greatest consumer interest.

Beyond simply satisfying taste and other preferences, Canadians report that they make food choices:

- to enhance general well-being (65%)
- to gain desirable nutrients (63%)
- to contribute to weight control (44%)
- for medical purposes (42%)
- to enhance resistance to illness (42%).

When asked about links between diet and health, most Canadians (92%) state that foods can lower disease risks and promote health. A high proportion (85%) also believes appropriate food choices can help reduce reliance on medications.

An Ipsos Reid (2005) survey on natural health products found that 71% of Canadians said they regularly take natural health products such as herbal medicines, vitamins, minerals, amino acids, homeopathic medicines, flower essences, antioxidants, glandular extracts and enzymes. Seventy-seven percent said they use these products to maintain and promote health and 68% said to treat illness.

Other studies have examined the credibility consumers attribute to various sources of health information. An Environics (2004) poll compared changes in perceived credibility between 2000 and 2004, finding credibility of dieticians and nutritionists increased to 75% from 70%, credibility of Internet sources increased from 21% to 35% and the credibility of health associations decreased from 82% to 62%. According to the most recent Statistics Canada survey of Canadians' Internet use, 68% of Canadians use the Internet, most on a daily basis (Statistics Canada, 2005). Close to 60% of Canadians with Internet access visit websites offering health and medical information and the Internet is the second most common source of health information, ranking only behind health care practitioners. Over 40% of Internet users purchase goods and services online.

Where do consumers look for help in navigating the maze of information required to assess the relevance of new health technologies? Research regarding the knowledge, skills and training of various health professionals, as discussed in Chapter 8, raise interesting questions about the common advice for consumers to 'consult with your doctor' about nutritional advice and supplements. Physicians and pharmacists rate themselves among the least knowledgeable of all the health professions in regard to nutrition. Other professions rate dieticians/nutritionists and naturopaths as the most qualified to advise consumers about health-related dietary choices.

NUTRIGENOMICS FOCUS GROUPS

Focus groups, which are guided interviews in a group session, can be used to help test theories, gather perceptions and discuss the potential for benefit and harms of new technologies. Between November 22 and November 29, 2007, 12 focus groups were conducted with members of the general public in five cities (Vancouver, Edmonton, Toronto, Montreal and Halifax). Four focus groups in two cities (Vancouver and Toronto) with health care professionals (doctors and pharmacists in two groups and naturopaths, dieticians and nutritionists in the other two groups).

To examine multiple layers of perceptions, several information sources about nutrigenomics were used to elicit responses. These included:

- pre-testing for existing knowledge and then examining general questions about participants' knowledge of and interest in the links between genetics and nutrition
- testing responses to a media article on nutrigenomics
- testing responses to a mock website offering nutrigenomic services
- testing responses to a scientific fact sheet that summarizes the current state of the science of nutrigenomics.

Pre-test

The goal of any pre-test is to determine participants' latent understanding of a subject before they engage in discussion and are exposed to further information. Before each focus group, a written mini-questionnaire determined the consumers' and health professionals' perceptions of nutrigenomics. The pre-test revealed that most participants had never heard the term 'nutrigenomics' and only about one-third could provide any meaningful feedback when asked to define the term. Only a few provided an interpretation that included both nutrition and genetics. None had heard of any nutrigenomic services marketed to Canadians. These results did not differ greatly among the health care professionals, of whom only a few had heard of the term 'nutrigenomics'. These few were either dieticians or naturopaths. The physicians and pharmacists had never heard of the term but were able to intuit that the term refers to a link between diet or nutrition and genetics.

Media Article

To elicit responses to popular press reporting about nutrigenomics, focus group participants were asked to read an Associated Press newspaper article of 544 words that had been published in a Canadian newspaper (see Appendix 10.1 for a copy of the article). This article provided basic information about nutrigenomics in the context of a story of a registered dietician who had her DNA tested by a company and received personalized nutrition advice.

The participants' reaction to the article was mostly positive – reflecting a relatively upbeat tone in the article – and they reported that it sparked curiosity, interest and even a level of excitement in the subject. They said it made them think about the possibilities of personalized medicine and increased their knowledge. They said it provided perspective and clarity regarding the role genes play in their health and how diets and genes interact. After reading the article, most participants expressed a view that genetics and nutrition can have a significant impact on their lives and clarified for them how diets and genes interact.

Some participants reacted with guarded skepticism. A number of participants said the commercial or marketing aspect to nutrigenomics made them uneasy and that they resented being marketed to with this new technology. Some said that basing dietary advice on genetic testing represented, to them, a form of extreme preoccupation with health. Others said the price of the test mentioned in the article – under $100 – would make them question its accessibility or usefulness. That is, if the test is important, why is it so inexpensive? Some participants stated that they found the thrust of the newspaper article disturbing or disquieting and explained that they would worry about what might be found or detected through such testing.

Participants identified a number of issues that stood out for them as problematic. These included: the very preliminary nature of the science of nutrigenomics; the possibility of inaccurate DNA-based dietary advice; the fact that there were very few genes currently being tested; and the online marketing of services by companies primarily based in the USA. In terms of skepticism, the orthodox health professions – especially physicians and pharmacists – expressed the strongest reservations about nutrigenomic services. Their skepticism pointed to an apparent lack of proof or evidence of the effectiveness of these services, the sense that testing was limited, a lack of regulation and oversight of these services, the impersonal nature of the service, the overt marketing of nutritional supplements and the potential conflicts of interest, especially companies involved in both carrying out tests and selling supplements. Some expressed a view that they would have heard of nutrigenomics if the testing were proven with established benefits for consumers. Since they had not heard of nutrigenomics, they perceived that the area must be too novel to be useful. Some, however, considered nutrigenomics testing to be intriguing and suggested it could be beneficial. Despite the strong skepticism, most health care professionals strongly supported the idea that, as the technology becomes more widely available and known, professionals will need to know more about nutrigenomics to counsel patients. Without the knowledge, they may be caught off guard by patients who are interested in nutrigenetic testing or who have already purchased tests over the Internet and seek help with interpreting results and tailoring their diet to their genetics.

Website

Participants expressed more critical or skeptical perspectives after viewing a mock website modeled on examples of firms that currently offer nutrigenomic services via the Internet (see Appendix 10.2). This heightened level of skepticism was based principally on the site's marketing or commercial aspects and the participants' feeling that important health services were being tainted by commercialism. There was a suspicion that those marketing nutrigenomics products and services would put profits before health. Many participants reacted most negatively to the marketing of nutritional supplements in conjunction with genetic tests. Finally, some participants doubted the credibility or reliability of Internet-based services and said they would feel more comfortable accessing nutrigenomic services through a clinic or laboratory in their community.

Scientific Fact Sheet on Nutrigenomics

After viewing and discussing the newspaper article and the mock website, the focus groups received a scientific fact sheet prepared by the study authors and reviewed by a leading Canadian nutrigenomic scientist (see Appendix 10.3). The fact sheet was intended to provide a summary of the state of the science of nutrigenomics and a brief overview of commonly cited advantages and drawbacks of nutrigenomics.

Both consumers and health professionals felt that the fact sheet offered a balanced overview of nutrigenomics, appreciating that the benefits of nutrigenetic testing are uncertain at present and that the ambiguity and lack of definitive proof means that any discussion about the subject will be about potential benefits and potential risks. Focus group participants noted that the science of nutrigenomics has a strong possibility of improving one's health, could deliver a better and more tailored approach to nutrition, could detect or diagnose predispositions and help design preventative measures. They noted that such knowledge could promote a greater interest in health and diet and may reduce health care costs. Most participants said that a tailored approach to nutrition and health that addresses disease predispositions and supplies preventative options would be a good thing.

Understanding perceptions of risks related to nutrigenomics is important to help address such concerns in public communications about nutrigenomics. In the focus groups, the most commonly mentioned risk or harm related to nutrigenomics was the possibility of a misdiagnosis or a mix-up of test results leading to faulty or mistaken advice that could adversely affect a person's health. Others criticized the testing, saying that it should not be considered a 'magic bullet', and suggested there is a risk that people could become over-reliant or overly optimistic about nutrigenomics, especially if they expect it to be the answer to all their health problems. Another drawback frequently identified was the potential for ignoring more traditional approaches to health care, including regular check-ups. Some suggested that results of testing might not be clear or conclusive and that this ambiguity might foster unnecessary worry or concern among individuals. Many participants felt that people might become rigidly fixated on a specific diet or health if nutrigenetic test results show that their bodies harbor some vitamin or nutrient deficiency or disease predisposition. Some said that people might overcompensate for a particular deficiency in their diet or disease predisposition and thus cause different nutrient deficiencies or other health problems.

Since nutrigenetic tests and services are offered through private companies, many participants noted that there may be potential (however unlikely) issues related to consumer fraud, with companies delivering phony test results and recommending unneeded and costly supplements. Many participants expressed concerns about privacy of their personal information and DNA samples, wondering what would be done with their test results or who would have access to them. It was noted that this concern had special relevance as to whether insurance companies might be able to access this information to deny coverage based on test results.

The perceived lack of significance of the testing was frequently noted in much of the focus group discussion. This may be due to the fact only five examples of food–gene interactions were given in the factsheet and people are aware that their bodies have thousands of genes. Participants wondered how significant this testing would be if it covered so few of the thousands of genes that make up the human body and people questioned the possibility that they might have a predisposition or condition that testing would not uncover.

Many participants felt that nutrigenetic testing could have a negative impact on health by promoting anxiety among those who undergo testing. Other concerns were related to the potential for inconclusive results, delays in getting results, over-interpretation of results (i.e. reading things in the results that may not be valid) and identifying conditions one is unable to do anything about.

Overall, people reported that the perceived benefits of testing outweighed the risks even though there is uncertainty about both. Many participants said the newspaper article raised their interest and motivated them to try to seek out more information about nutrigenomics. Even though many of the participants said they were interested in the topic, this might be due to the 'focus group effect' where participants, knowing the interests of the researchers, try to give researchers what participants think they want to hear (i.e. the more socially desirable answer). Some participants admitted they were not interested in nutrigenomics or had a very low level of interest in the topic, saying that it seemed too complicated or difficult to understand. Many people questioned the technology and said they would want to know the evidence behind the claims, especially made by companies offering nutrigenetic tests and services, whether the tests were certified or regulated, how accurate they are, who does the testing, whether it is available in Canada and what it would cost.

There were a number of areas where the responses of the focus group participants and the participating health professionals were unambiguous. These include:

Need for information and regulation: there was unanimity that nutrigenomics testing should not be left exclusively to the marketplace and that a government body – in Canada, the federal health department or the Public Health Agency of Canada – should be involved in informing consumers about the benefits and harms of such testing and support appropriate counseling of patients by providing information resources about nutrigenomics for the public and health care professionals. Participants suggested they would like to know information about the current state of the science of nutrigenomics, details about companies that market tests, including whether they have been the subject of complaint or investigation for fraudulent practices. They also proposed that a regulatory body should oversee professionals who provide nutrigenomic services.

Trusted health sources: many participants identified family physicians, dieticians and nutritionists as most credible and trustworthy sources of health advice and information. Yet, in the context of nutrigenomics, participants expressed the feeling that these health care professionals did not yet have expertise in this new area. This revealed an important quandary for consumers who would like to seek advice from a trusted source, but doubt the current capacity of that source to provide adequate counseling. Health care professionals said they would mostly refer patients to geneticists, nutritionists and dieticians as well as researchers or professors at universities or medical schools for further information about nutrigenomics.

Public over private sources: people expressed uncertainty in terms of where they would go for further information about nutrigenomics but favored public rather than private sources. In terms of trusted sources, most suggested their physicians' offices, clinics and health care facilities, as well as government/regulatory agencies such as Health Canada, other public health agencies, the US Food and Drug Administration, or simply government, in general.

CONCLUSIONS: KEY BENEFIT AND RISK INFORMATION TO INFORM THE PUBLIC AND HEALTH CARE PROFESSIONALS

Both consumers and health care professionals identified a wide spectrum of information they felt was important for Canadians to know to make informed decisions about nutrigenomics. This included:

- a balanced assessment of the advantages and drawbacks of nutrigenomics based on current scientific knowledge
- an identification of reliable sources of information about nutrigenomics
- a 'points to consider' document outlining factors pertinent to making a decision about nutrigenetic testing
- a list of nutrigenomics companies with contact information
- a statement on how nutrigenomic service providers are currently regulated
- an explanation of whether nutrigenetic testing is an insured service under public health care programs
- a statement of Health Canada's position regarding nutrigenomics (e.g. does Health Canada endorse nutrigenomics and why or why not?)
- a clear statement of privacy and confidentiality considerations (e.g. how are genetic samples and personal information protected, who has access to information and what rights do consumers have?).

When asked, people want to know, in simple and clear terms, the 'pros and cons' of nutrigenomics and some demand proof or evidence of examples of the success of nutrigenomics. They want to know who endorses nutrigenomics, how reliable it is, who does the testing and where the testing is done. They want to know what nutrigenomic tests can test for, if and how services are regulated and who is responsible for maintaining the privacy of information collected. They want to know how much the technology costs and whether it is publicly insured. Consumers and health care professionals believe that many actors have a supporting role to play in helping provide information about nutrigenomic testing, especially Health Canada, the community of professionals involved in providing health care, associations representing health care professionals, university researchers and medical schools.

This study determined that public awareness of nutrigenomics is extremely low at the moment – indeed, almost non-existent. This precludes saying anything substantive about consumers' perceptions of the risks associated with this technology, there is value in establishing a baseline understanding of consumer perceptions. Seeking to understand current awareness and attitudes about nutrigenomics and identify knowledge gaps is useful for two reasons:

1. to ensure the field develops and is communicated in a way meaningful to consumers and professionals
2. to target strategic investments to close knowledge gaps.

It would also seem that this early work points out the fact that there is sufficient time for decision-makers at all levels to think clearly and thoroughly about how the benefits and the risks of the science are to be optimally communicated to the public.

This research has established a baseline measure of the type of information that consumers and health professionals say they want to know, as well as suggested ways best to inform Canadians about nutrigenomics and the kind of expectations they have of government to regulate or oversee nutrigenomics services. There was a strong consensus that nutrigenomics testing should not be left to the marketplace. Health professionals want to be able to answer their patients' questions and concerns and therefore have a strong desire to be supported by government in helping deal with patients' information requests regarding nutrigenomics. In particular, health professionals stated that Health Canada and/or the Public Health Agency of Canada should:

- provide information to health care professionals and professional associations about the government's position on nutrigenomics, create pamphlets or brochures for distribution to the general public through health professionals' offices/clinics, as well as provide a list or registry of nutrigenomics companies and make available any information about regulatory action taken against those companies
- sponsor or fund independent research into nutrigenomics through universities or research institutions and fund public laboratories to carry out nutrigenomics testing
- provide support to naturopaths and alternative health care providers who provide patient care and counseling regarding nutrition and health

- consider creating an oversight body (such as a College of Nutrigenomics Practitioners) to ensure professional standards and credentials
- create an advisory team (comprised of physicians, pharmacists, dieticians, nutritionists and naturopaths) to make recommendations regarding regulation of nutrigenomics.

As the science continues to evolve, the future of nutrigenomics will depend to some extent on acceptance by the public and health professions. Our research indicates that both groups expect government to be proactive in providing information sources about nutrigenomics. In addition, they want to see that government regulators, commercial interests and others involved in the field adopt policies and practices that minimize risks and maximize benefits.

APPENDIX 10.1

Diet: Scientists look to DNA for personalized advice on nutrition
Author: Malcolm Ritter AP Science Writer
Date: October 13, 2005
Publication: Associated Press Archive

New York (AP) – As a registered dietitian, Ruth DeBusk has eaten a healthy diet for a long time. As a geneticist, she wondered if she could do better.

So earlier this year, she had her DNA tested by a company that gives personalized nutrition advice based on genetics. The results indicated she needed more folate.

So DeBusk doubled her minimum amount of folate, a B vitamin found in leafy greens and citrus.

'I'm more diligent about being sure that I get it every day if possible, because it really matters', said DeBusk, who has a private practice in Tallahassee, Florida, and has written a book on nutrition and genetics.

'I'll actually make an effort to drink a glass of orange juice or eat an extra big salad in the evening, being aware it hasn't been one of my better folate days'.

That's the way it's supposed to work in a field called nutritional genomics or nutrigenomics. The basic idea is this: there are genes that affect the risk of getting illnesses like heart disease, cancer, osteoporosis and diabetes, and the impact of those genes can be modified by what you eat. Everybody carries one version or another of each of those genes. So why not find out what gene versions you have and base dietary advice on that?

'Every time we go to the supermarket we're using educated guesses about what we should eat and what we shouldn't eat', says Raymond

Rodriguez, director of the National Center of Excellence for Nutritional Genomics at the University of California, Davis.

In the future, more of that guesswork may be replaced with accurate, personal DNA-based dietary advice, which Rodriguez says is 'rapidly emerging on the horizon'.

But that time isn't here yet, most experts say. Nutrigenomics is still in its infancy, with plenty to be learned, and it's not yet clear what role it may play in standard medical practice.

Most of the research targets heart disease and cancer, and scientists may be ready to deliver personalized diet recommendations in those areas within five years, said Jose Ordovas, director of the nutrition and genomics laboratory at the US Department of Agriculture Nutrition Research Center at Tufts University in Boston.

'We have scientific evidence that the concept is right, that we can provide something along those lines in the future', Ordovas said. 'We are not there yet.'

Right now people, wanting an idea of what they should be eating, can pay $99 US for a DNA test kit that will provide personalized diet advice for heart health, bone health, or any of three other areas.

It's from Sciona Inc., a small company based in Boulder, Colorado, that started offering DNA-based diet advice in 2001. The tests are available by mail order and on the Internet.

Sciona customers collect their own DNA with a cheek swab, complete a diet and lifestyle questionnaire and send it all in for analysis. Sciona encourages customers to review its advice with a doctor.

The company acknowledges that some scientists say it's too soon to offer such a service, but says its testing is based on solid research. Current testing focuses on 19 genes and the company is studying others, said Rosalynn Gill-Garrison, chief scientific officer and a company founder.

APPENDIX 10.2

Home Page for 'Nourigena Health'

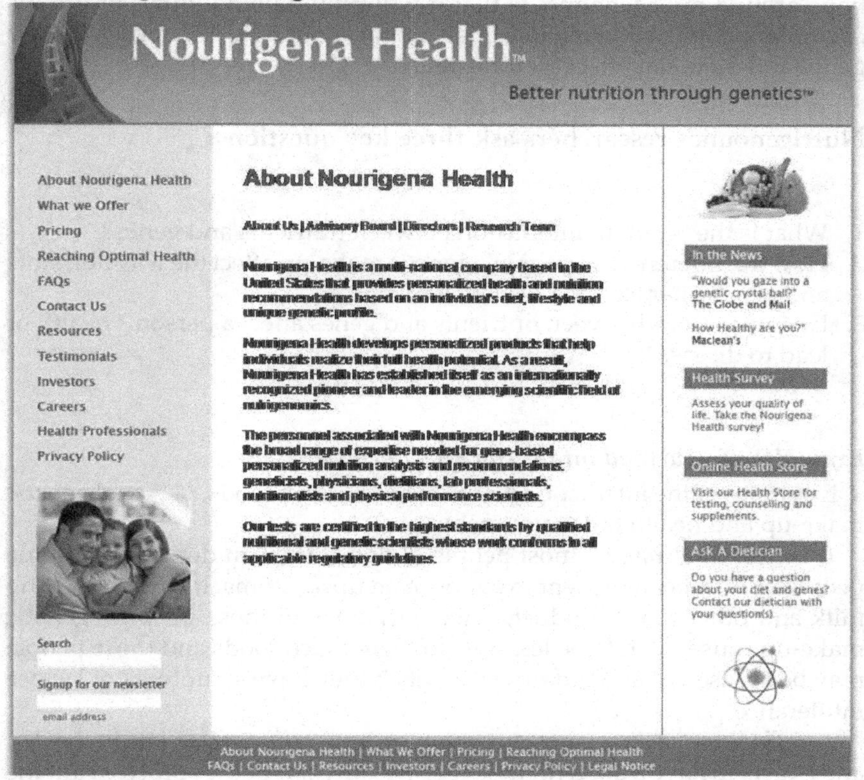

APPENDIX 10.3

Introduction

Nutrigenomics is a new and developing science that studies the interaction between the nutrients in our food and the genes in our bodies. Nutrients in food, such as proteins and vitamins, provide our bodies with the energy we need to live. Genes are the basic biological materials that provide instructions to build and maintain our bodies. Nutrigenomics examines how nutrients and genes interact, where our genes affect how our bodies use nutrients and where nutrients influence how genes work.

Studying nutrigenomics usually involves collecting a sample from a person (usually a cheek swab but, in some cases, through drawing blood) and looking at that individual's genetic make-up (this is a genetic test). Other personal information (such as gender, weight, age and other factors), plus an analysis of the nutrients normally consumed may also be examined. The hope around nutrigenomics is that it will provide individuals or groups of similar people with valuable information about how differences in their genes and diets can affect their health.

Nutrigenomics researchers ask three key questions:

1. What is the 'normal' interaction between nutrients and genes?
2. How do differences in people's genetic make-up affect the way nutrients and genes interact?
3. Do interactions between nutrients and genes affect a person's health or lead to disease?

Examples of how food intake can affect health

Examples of the interaction between nutrients in foods, personal genetic make-up and health outcomes include:

1. *Lactose intolerance* – most people of non-European descent and some people of European descent experience an upset stomach when digesting milk and other dairy products. Two variations in those peoples' genetic make-up cause them to be less able to digest such foods and those people may be advised to avoid dairy products if they have symptoms of lactose intolerance.

2. *Caffeine metabolism* – not everyone can have coffee or tea late in the day and expect to have a good night's sleep because our bodies process caffeine differently. There may be a genetic reason why caffeine is processed more quickly by some people than others and those who process caffeine more slowly may have a higher risk of heart disease and thus might wish to limit their daily caffeine intake.

3. *Folate supplements* – homocysteine is a chemical in our blood that is associated with heart disease. Some people may have a high homocysteine level because of their personal genetics. They may have a gene that slows the speed at which an enzyme can clear homocysteine from their body but, by taking folate supplements (also known as folic acid), these people can lower their homocysteine, thus reducing their risk of heart disease.

4. *Dietary fat and body weight* – humans consume different kinds of fats and those who get a significant amount of their daily energy from eating fatty foods are more likely to be overweight. However, there are some people with a slightly different genetic make-up that allows them to consume high

fat diets without gaining as much weight as expected. Since your genetic make-up can affect how your body processes certain types of fat, knowing those differences could help you steer clear of those kinds of fat and thus better manage your weight.

5. Diets and populations – in parts of Arizona and Mexico, there is a group of aboriginals called the Pima Indians. The Pima tend to stay closely-knit, especially in Arizona where they live on a reserve. Pima Indians who consume their traditional diet (high carbohydrate, low fat and low protein) tend to be slim, but those who consume a typical, modern diet are almost 20 times more likely to develop diabetes than the average American. This response to diet raises important questions about the difference in response to diet of people who share common ancestry.

Nutrigenomics: Some Issues

There is debate in the scientific community around the new and developing field of nutrigenomics. One perspective is that:

- knowing one's personal genetic make-up could be a foundation for personalized nutrition and better health
- tailoring one's diet to their specific genetic make-up could help prevent, reduce or delay the onset of certain diseases; and
- nutrigenomics could promote an increased understanding of how nutrition influences genes and the information could be used to prevent the development of disease (e.g. chronic diet-related diseases such as obesity and diabetes).
- Another perspective is that:
- it may be too early to apply nutrigenomic science at this relatively early stage of its development
- nutrigenomic (genetic) test results are rarely 'black and white', in most cases, disease is caused by a variety of interactions between genes and many environmental factors
- some companies marketing nutrigenomic tests may overstate the benefits of the test results/information they provide to customers
- at home genetic tests have not yet been evaluated by government regulators; and
- Internet marketing of these tests may be:
  - making predictions that are not yet accurate and/or medically proven; and
  - used as a means of marketing costly but possibly unnecessary 'personalized' dietary supplements.

References

Decima Research (2006). Demand for health products supporting health and wellness 2006: final report. Prepared for Agriculture and Agri-Food Canada. Available at: http://www4.agr.gc.ca/resources/prod/doc/misb/fb-ba/nutra/deman/pdf/decima_2006_e.pdf.

Environics (2004). Health perceptions around the globe. Available at: http://erg.environics.net/.

Klein, W.M.P. and Stefanek, M.E. (2007). Cancer risk elicitation and communication. *CA Cancer J Clin* 57:147–67.

Ipsos Reid (2005). Baseline natural health products survey among consumers: final report. Prepared for Health Canada. Available at: http://www.hc-sc.gc.ca/dhp-mps/alt_formats/hpfb-dgpsa/pdf/pubs/eng_cons_survey-eng.pdf.

Statistics Canada (2005). Canadian Internet use survey. Available at: http://www.statcan.ca/Daily/English/060815/d060815b.htm.

CHAPTER

11

Pitching Products, Pitching Ethics: Selling Nutrigenetic Tests as Lifestyle or Medicine

Paula Saukko

OUTLINE

Nutrition and Genomics
ISBN: 978-0-12-374125-7

SUMMARY

This chapter argues that ethical and regulatory debates around commercial nutrigenetic testing are pushing the technology in two contradictory directions: becoming more associated with lifestyle or becoming more medical. The chapter will briefly analyze the web pages of select nutrigenetic testing companies as well as genetic test results purchased from two companies, the US-based Sciona and the UK-based Genetic Health. The web pages frame their products as lifestyle or medical through the use of words (i.e. discussing wellness or illness), colors and images. The Sciona test report compiled the results of various genetic tests into broad statements about health priorities and recommended eating a generally healthy diet (low in saturated fats, rich in antioxidants), taking exercise and multivitamins. The Genetic Health report provided advice based on the results of individual gene tests, including recommending against eating a Mediterranean diet, monitoring for breast and prostate cancer and giving advice on hormone replacement therapy. The two tests reports illustrate a more lifestyle oriented approach (Sciona) and a more medical strategy (Genetic Health). In the short term, in a situation where the technology of nutrigenetic testing is not established and validated, providing general advice may be less problematic than specific recommendations. It is difficult to predict how nutrigenetic testing will develop as a science and a commercial enterprise in the long term. If clinically useful nutrigenetic tests were to become established in the future, a more medical approach may be warranted, but it would need to balance effective prevention and unnecessary medicalization.

INTRODUCTION

Commercial nutrigenetic tests – which provide DNA-based dietary advice and are typically sold direct to consumers online – have been accused of misleading the public (e.g. Haga et al., 2003). The controversy focuses on two issues.

First, it has been argued that people may learn about their genetic predisposition to serious illness via an online test without appropriate counseling and support. The example most often cited in this context is the APOE E4 mutation. APOE E4 is associated with an individual's increased susceptibility to heart disease. For these individuals, a diet low in saturated fats and high in omega 3 fatty acids would be advisable. APOE E4 is also associated with increased predisposition for Alzheimer's disease, for which there are no preventive strategies (Ordovas, 2004). Thus, there is a possibility that an individual will learn about a susceptibility to serious neurodegenerative disease, for which no cure or prevention exists, as a side result of nutrigenetic testing.

Second, it has been argued that the information provided by commercial nutrigenetic tests is spurious in that the scientific evidence to support the tests is too weak and the advice given too general. It has been stated that nutrigenetic tests provide little added value beyond reiterating obvious advice that people should eat healthily, exercise plenty and stop smoking, which a recent consumer report called 'run-of-the-mill advice' (Consumer Reports, 2006).

In this chapter, it is argued that the ethical, regulatory and popular debates around nutrigenetic tests are pushing the technology in two, contradictory directions. First, the argument that nutrigenetic tests may provide harmful or serious health information for people without proper medical guidance invites the companies to frame and develop their products in a way that render them more ambiguous and less serious – more 'lifestyle' than medicine. Second, the claims that nutrigenetic tests are not based on sufficient evidence and tend to provide generic health advice encourages the companies to frame and develop their products in a way that renders them more specific and serious – more 'medicine' than lifestyle.

An analysis of the websites of eight companies selling nutrigenetic tests online indicates that the demands of being both less and more serious often translate into contradictions in the marketing of the tests. Different companies are also choosing different paths in marketing and developing their products more as lifestyle or more as medicine. Companies respond in this way because they are anticipating ethical and regulatory responses. Hedgecoe and Martin (2003) and Hedgecoe (2004) have argued that ethical and regulatory concerns do not simply respond to technological developments but actively shape the development of new technologies. This is also happening in nutrigenetic testing. This chapter charts the way in which nutrigenetic tests negotiate the pull between medicine and lifestyle and discusses its potential social implications.

BETWEEN HYPE AND CONCERN

Brown and Michael (2003) have coined the term 'sociology of expectations' to refer to research on how scientists and entrepreneurs create expectations or 'hype' about a new technology in a bid to make it into a reality. Rather than branding such expectations as misinformation, Brown, like others in science and technology studies (e.g. Rajan, 2006), note that the hype has highly productive dimensions. Scientists need to create a momentum for a new science, such as nutrigenomics or stem cells, if they want to attract funding for their research. Further, biotechnology companies, which seek to translate the research into practice and commercial activities, need to create expectations to garner early stage investment capital and then consumer interest.

Expectations are not only positive. New scientific innovations and technologies also spark scenarios of fear and risk, as is most clearly evident, for example, in debates over human cloning and genetically modified crops. Thus, hype surrounding new technologies not only mobilizes proponents but it can also galvanize opponents and critics. In response to this, public officials have sought to design processes and principles, such as the 'precautionary principle' in risk analysis, which encourage the engagement of diverse stakeholders prior to approving technological projects and innovations. These processes enhance the potential for public acceptance of new technologies.

Hedgecoe and Martin (2003) observe that social scientists, bioethicists and regulators often see themselves as responding to new technological developments and the concerns they raise. They argue that, in fact, ethical and legal concerns may shape the way a technology is developed, sometimes through a series of negotiations that take place before a new scientific technology is translated into clinical practice or commercial product. Thus, the technology of genetic testing *per se* does not determine its applications, which are shaped by varied interests, including ethical concerns, in a specific historical, social context. In health and medicine, concerns about new technologies are often articulated in terms of ethical issues, which may concern individuals' safety or more fundamental ethical principles, such as the integrity of nature. The latter is particularly the case, for example, in concerns raised about cloning.

An example of the way in which ethics shapes scientific applications is provided by pharmacogenetics. The technology has been envisioned in two ways. First, pharmacogenetic testing has been defined as focused on drug metabolism and being independent of disease-causing genes. Second, pharmacogenetics has been associated with genes conveying susceptibility for disease and leading to reclassification of disease, for example, various genotypic subtypes of diabetes, which respond to different drugs (Hedgecoe, 2004). Hedgecoe and Martin (2003) argue that the first vision became the accepted one, mainly due to ethical and legal reasons. What happened was that bioethical discourse distinguished between 'purely' pharmacogenetic tests and genetic tests for genes that cause disease. The former was thought to pose less significant ethical, legal and social risks. In anticipation of greater public and clinical acceptance of pharmacogenetics, tests were developed as simply detecting susceptibility for side-effects of drugs. The idea of pharmacogenetic tests for identifying variants of the disease itself, which would respond differently to drugs, was abandoned. The issue, however, remains latent as many pharmacogenetic tests do indirectly detect disease susceptibility as well.

NUTRIGENETICS: BORDERLINE SCIENCE/PRACTICE

Nutrigenetic tests provide another timely case for examining the process of translating frontline science into a commercial product. Several features make nutrigenetic tests an interesting case study. First, nutritional science has traditionally been associated with the less prestigious, female discipline of home economics, and its marriage with the hard science of genetics has produced a contradictory, both high tech and low tech, disciplinary identity. Second, nutrigenetics is not only rendered contradictory by its interdisciplinarity but also by wider, historical trends that have confounded the identity of medicine as it has shifted from curative to preventative medicine. Infectious diseases now take a back seat to diseases associated with contemporary lifestyle, such as obesity and type 2 diabetes. Practices traditionally considered falling outside medicine, from acupuncture and homeopathy to exercise, are also being integrated into clinical practice, in a bid, particularly in family medicine, to provide 'whole person care'. At the same time, the complementary practices are being mainstreamed by attempts to prove their effectiveness in clinical trials (Ernst, 2006). But practices and products traditionally considered medical are also being made available in less specialized settings, so that, for example, cholesterol testing as well as over-the-counter cholesterol lowering drugs (statins) are available in some countries, such as the UK, in pharmacies. These developments have sparked accusations about medicalization of everyday life (Lippman, 1998) as well as trivialization of medicine (Haga et al., 2003). Online nutrigenetic tests belong to this emerging array of products and services and present themselves with an unclear identity.

Third, the development of nutrigenetic tests has not passed unnoticed by opponents and regulators. Due to the novelty of the technology and the fact that it does not clearly belong in any previous regulatory categories, there are currently few specific restrictions on the sale of the tests online (Hogarth et al., 2007). Attempts have been made to classify genetic tests into serious or 'high impact' and less serious or low/intermediate impact both in the USA (Secretary's Advisory Committee on Genetic Testing, 2001) and UK (Human Genetics Commission, 2002, 2007). These categorization attempts have, however, failed to provide a regulatory environment in which nutrigenetic tests are regulated directly. Currently, genetic tests sold directly to the public are classified as 'not serious' in the European Union legislation and are, therefore, exempt from pre-market review (Hogarth et al., 2007). Attempts have focused on regulating the marketing of nutrigenetic tests via consumer and advertising regulation by identifying possible false health claims made by the companies (Human Genetics Commission, 2007). Companies selling the tests respond to the novel and uncertain regulatory situation by, for example, modifying their practices so as not to appear to be selling the most serious or high impact tests, or to appear to be making false claims.

The situation now is that companies pitch their products not only in terms of a market niche but also in terms of a regulatory or ethical niche. Thus, the development and marketing of nutrigenetic tests is an example of the process of ethical and legal practices shaping new technologies, as observed by Hedgecoe and Martin (2003). In this case, the pressure on companies leads to a decision about where to market the nutrigenetic tests along a continuum between lifestyle tests and medical tests. During fieldwork, an industry representative stated that the situation is such that companies need to decide whether they would start to sell nutrigenetic tests in association with 'diet drinks and skin creams', which nobody seriously believes to be effective, or in terms of cardiovascular products, which would need to be backed up by scientific evidence. In an April 13, 2007 interview, the chief science officer of Sciona, Rosalynn Gill, referred to a similar situation by saying that the company cannot be 'perceived as being involved in the practice of medicine'. This was in response to a question of why Sciona does not test for the factor V Leiden gene variant, associated with a more significant risk of deep vein thrombosis, but instead tests for the MTHFR gene variant considered a softer marker for cardiovascular disease.

EMPIRICAL STUDY

To explore how companies selling nutrigenetic tests directly to the public market their products, we analyzed the web pages of eight companies in the USA and the UK: Interleukin Genetics (USA), Genelex (USA), Genovations (USA), Genetic Health (UK), Market America (USA), Nutrigen (USA), Sciona (USA, previously UK) and Suracell (USA). The sample consists of companies we could find selling nutrigenetic tests online in summer 2007; the sample is not complete but captures the majority of companies operating at that time. Testing companies come and go and, in some cases, are also complexly interconnected so that the same test may be available through different providers under a different name, so they are somewhat difficult to identify. We not only downloaded the current websites but also analyzed older websites that could be traced using the Wayback Machine. Not all companies' pages could be found on Wayback Machine but, for instance, Sciona's pages were available from 2001 when the company started operating. We also purchased a nutrigenetic test from Sciona (USA) and Genetic Health (UK) in 2007 and analyzed the advice received.

Drawing on the constant comparative method (Glaser, 1965), an analysis of key themes emerging from the websites was conducted with the aid of NVivo qualitative software. The way in which the companies teetered between marketing their products and giving advice in terms of lifestyle and medicine emerged as a major theme. To illustrate how the companies pitch their products in relation to this distinction, four aspects of marketing and advice are explored:

1. instances where nutrigenetic testing companies associated their products with wellness or lifestyle
2. instances where they associated their products with illness or medicine
3. instances where the advice provided by the companies was expressed in terms of lifestyle
4. instances where the advice provided by the companies was expressed in medical terms.

Pitching wellness

Genetics is traditionally associated with serious, often incurable, single-gene disorders, such as Huntington's disease. Nearly all companies underlined that the genetic tests they were offering were different from tests for single gene disorders. For example, on its website, Market America characterizes its testing as follows:

> *Can you tell me if I carry the genes for a serious illness?*
>
> No. The Gene SNP DNA Screening Analysis is *not a test for inherited disorders or inherited predisposition to disease.* We do not screen for disorders caused by a defect in a single gene, such as Huntington's disease, cystic fibrosis or sickle cell anemia. Our screening process is relevant only for the much more common situation: the presence of gene variations that influence a person's *ability to derive maximal benefit from diets and lifestyle practices* recommended by current medical research (Market America, 2007, emphasis added).

This quotation indicates the companies often emphasize that their products help maximize health rather than diagnose or prevent illness. This is often expressed in terms of helping individuals 'optimize' health or wellness:

> *Optimize the health of your skin* and bones; heart and mind *by optimizing your personal diet* and supplement intake. Genetic testing combined with a lifestyle assessment, provide you with a scientifically based, personal blueprint for *optimizing health* (Genelex, 2006, italics added).

Rose (2007) has discussed ethical debates around medical products which are claimed not only to cure illness, but to enhance performance or moods, particularly in relation to psychopharmaceuticals. This kind of enhancement has been accused of going beyond what is acceptable in medicine (i.e. to restore normalcy) and tampering with human nature to create an artificially modified happy or healthy self (see e.g. Kramer, 1993 for a discussion). Nutrigenetic tests have not been accused of fostering inauthentic super-humanity, although this might become an issue in the future, particularly with respect to sports nutrition and genetics. Rather, the companies seek to market their products in association with a positive concept of wellness or vitality rather than a depressing or anxiety-provoking illness.

As such, the testing companies become part of the burgeoning wellness industry, as embodied in different kinds of wellness centers, which may offer exercise and yoga classes, massages, herbal and homeopathic treatments alongside traditional vitamin supplements and conventional drugs. The wellness services and products do not typically seek to cure illness, but to enhance personal well-being and professional performance by providing relief from stress and tension and boosting the immune system, energy levels or moods. Nutrigenetic tests retailed directly to the customer are often sold through such wellness centers or 'holistic' pharmacies. Indicative of this development is the transformation of the London Harley Street (where Genetic Health offers nutrigenetic testing) from being renowned for its private doctors for the rich to a new hotspot for complementary medicine.

Wellness products and services often market themselves as more 'natural' than conventional medicine. Nutrigenetic tests are no exception to this and their marketing often emphasizes that the advice is in harmony with 'nature' (i.e. DNA). An example can be found in the description for anti-aging nutritional supplements in one of the websites:

> The nutritional supplements ... combine dozens of high-density nutrition ingredients with a proprietary extract from *the rainforest botanical* Uncaria Tomentosa, called CAEs, that is perhaps the single most proactive means of promoting the *body's own natural DNA* care and maintenance abilities (Suracell, 2004, emphasis added).

Lury (2000) has explored the contradictions embedded in the trend in consumer culture to commodify nature and sell it to customers as the 'real thing'. Yet concerns about nutrigenetic tests have usually not focused on their alleged naturalness. Rather, the tests have been criticized for being nonsense (Haga et al., 2003).

Thus, companies face the challenge of marketing their products both in terms of promoting natural wellness and in terms of embodying state-of-the-art science. The web pages illustrate this tension. All eight companies analyzed in this study make references to high science and published literature, as well as their laboratory accreditations, to signal seriousness and credibility alongside naturalness and optimal wellness. An example is Sciona's mission statement in its 2007 web page:

> Sciona has built a business around the application of *evidence-based* genetic information for the development of personalized products that help individuals optimize their

> health. . . . Its laboratory services are *certified under the Clinical Laboratory Improvement Amendments* passed by the United States Congress in 1988 to establish *quality* standards for laboratories and to ensure *accuracy*, reliability, and timeless [sic] of laboratory tests (Sciona, 2007, emphasis added).

The tensions are also articulated in the symbols used on the pages. The companies' web pages frequently present images of 'science', such as various stylized forms of the double helix, a man in a white laboratory coat peering into a microscope (Sciona), a male doctor reading notes (Nutrigen) or images of stethoscopes (Genetic Health). The color schemes of the web pages of some companies contain white, gray and dark red (Genovations) or blue and white (Genetic Health, doctors's area), traditional colors of medicine and officialdom. Other pages contain images of fresh, gleaming citrus fruits (Genetic Health, front page) or sweet peppers (Sciona) and images of semi-nude women in stark light or dressed in white (Suracell) or a female child in a snow-like environment (Interleukin) to signal purity and natural essence. The color schemes of web pages may also contain white and green, sometimes splashed with yellow, symbolizing vitality and nature (Nutrigen, Genetic Health, nutrition gene page). The images, colors and language on the pages seek to create an impression of both official science and natural wellness.

Pitching medicine

All testing companies refer to diseases, such as heart disease and osteoporosis, on their websites because their tests seek to identify disease predispositions and provide prevention-oriented advice. Some of the companies marketing nutrigenetic tests had adopted rhetoric and practices that position them closer to mainstream medicine than the others. An example is the introductory page of Genovations:

> Genovations^{TM} . . . allows a physician to evaluate each patient's unique genetic predispositions, then develop and implement a carefully targeted, customized *intervention plan even before pre-disease* imbalances are manifest. Virtually all of the most *pervasive, disabling, and deadly degenerative diseases of our time, including heart disease, adult-onset diabetes, cancer, and senile dementia*, are believed to develop from an ongoing interaction between genetic and environmental factors (Genovations, 2007, emphasis added).

This company uses clinical language, such as 'implementing' and 'intervention' and addresses customers as physicians and patients, symbolic gestures which all signal that the company seeks to associate itself with the sphere of official medicine. Genovations also uses more explicitly disease-oriented language than most other companies, mentioning 'pre-disease states'. Rather than discuss enhancing wellness, the company uses stronger language in describing potential illnesses, such as 'disabling', 'deadly' and 'degenerative'.

A UK company, Genetic Health, also frames its tests in a more medical manner on its web pages. For example, it explains the information its tests of single nucleotide polymorphisms (SNP) provide:

SNP analysis aims to answer some of the following questions:

- What is my individual risk of developing a particular disease?
- Am I more susceptible to environmental influences that can cause disease?
- How will my body react to any drugs I am being prescribed?
- How successful is the treatment regimen I am undergoing likely to be? (Genetic Health, 2007)

The answers, and the language in which they are framed, is again more resonant with clinical language than in the case of some of the other companies. The paragraph uses the words 'individual risk' and 'treatment regime', which refer both to clinical practices of individual risk estimation and treatments, as well as indicates that the tests are managed by a physician. Genetic Health also includes more potent or controversial alleles in its panels than, for example, Sciona. It offers a test panel with the APOE E4 allele, which is associated not only with heart disease but also with Alzheimer's disease. Another 'premium' panel includes the test for factor V Leiden, which is associated with an increased risk of deep vein thrombosis and is also offered through the UK National Health Services (NHS).

Lifestyle advice

The controversy over nutrigenetic tests concerns not only how companies market the tests, but also the kind of advice they offer people and the alleged effects of this advice. To assess directly the advice companies give to consumers, we purchased a nutrigenetic test online from Sciona (USA) and Genetic Health (UK) in 2007. In both cases, we sent a mouth swab of saliva (for Sciona we also sent a lifestyle questionnaire) and received the report in the mail. Although the two companies share many features, as indicated above, Sciona pitched its tests more as lifestyle than medicine, while Genetic Health, which had operated less than a year, gave a more medical appearance. This was also evident in the advice they provided.

Sciona tested for 24 gene variants covering five areas (bone health, heart health, detoxification, inflammation and insulin sensitivity). The company tested for 13 common variants associated with 'heart health' and presented the results of the tests together. The test we purchased indicated that five of the genes associated with heart health did not have 'an impact', whereas eight of them did. Thus, since each gene variant tested is common (present in 10–60% of the population, according to the report), all individuals end up having some of the risk genes, while not having others. So, everyone is a bit at risk to warrant some advice. The results of the heart health tests were reported in a general manner, stating 'you have variations' in the vitamin B metabolism/inflammation response/antioxidant defense gene. None of the summaries for the five areas covered by the genetic test indicated that the individual tested faced greatly increased risk of any condition, nor did they indicate that the individual was not at risk or protected against any condition.

The company mediated this vagueness by presenting a list of 'priorities' and various charts seeking to communicate levels of risk and action, based both on gene and lifestyle assessments. The test we purchased, which was the basic Cellf test covering five areas, indicated 'heart health' as first priority, followed by bone health and insulin sensitivity (second priority) and antioxidant/detoxification (third priority). Of the areas included in the report, inflammation was not listed as a priority. The relative importance of the diverse areas was further conveyed by charts called 'action plans', which identified a 'you are here' and 'your goal' place in relation to both 'gene assessment' and 'your lifestyle'. The chart was colored, ranging from red (action required) to yellow (optimal health). The chart resembled the traditional 'traffic lights' chart. For example, a clinical assessment of heart disease risk communicates information about an individual as green (low risk), yellow (moderate risk) and red (high risk) indicating specific levels of absolute risk. The Sciona action plans, however, did not identify numerical levels of increased risk and the 'you are here' points in all the charts for the five different areas ended up in the orange (most of the chart was orange). Thus, the charts sought to convey a sense of estimation/calculation without really saying more than that everyone is a bit at risk of common illnesses.

The recommendations made based on the gene and lifestyle assessment were also rather general. For example, the summary report for the antioxidant/detoxification results stated: 'You have variations in your genes important for antioxidant defenses which may lead to less efficient removal of free radical damage from your body'. The recommendation was to 'increase your consumption of cruciferous vegetables and foods high in antioxidants'. The summary report for heart health stated, for example: 'You have a variation in your cholesterol and triglyceride metabolizing genes'. The recommendation was 'to pay particular attention to the levels of unsaturated and saturated fats in your diet'.

The dietary recommendations were also conveyed in terms of charts, which mainly indicated current intake of vitamins (the lifestyle questionnaire submitted together with the mouth swab for the genetic test asked detailed questions about intake of food and vitamin supplements) and 'your goal'. The recommendations for all the areas ended up suggesting increased intake of most vitamins (vitamin B, omega 3 fatty acids, etc.), as well as recommending less saturated fats, more physical exercise and intake of, for example, cruciferous vegetables against 'current' levels estimated from the lifestyle questionnaire. Again, these charts give an impression of personalization while containing general advice about the importance of eating vegetables, eating less saturated fats, exercising and taking a multivitamin.

Sciona seemed to play it safe and not make too specific recommendations or risk predictions that would be difficult to defend in the context of an emerging science. While the generality of the advice was safer from accusations of being wrong and harmful, it made the company liable to criticisms that it was selling generic advice as genotype specific.

More medical advice

Genetic Health's nutrigenetic test included 16 gene variants in the areas of 'metabolism', 'detoxification', and 'inflammation'. Some of the alleles tested (MTHFR, GSTM1 and IL6) were the same as in Sciona's panel, others (e.g. CYP family of detoxification genes) were different. While Sciona combined the analysis of different gene variants for, for example, detoxification, the Genetic Health's report gave separate advice on each variant. It also indicated when gene variants were having a 'protective' effect on health, as opposed to simply reporting on the deleterious effects of the variant.

The advice in the report was more specific than the advice in the Sciona report and touched on issues that could be considered medical. For example, the test for an APOA1 variant concluded: 'Your APOA1 gene shows no polymorphism. Your HDL rates are rather low. Especially for you, an ample dietary take of polyunsaturated fatty acids (abbreviated PUFA's) *is not good*. The Mediterranean diet, generally considered to be healthy, is *not* very suitable for you'. This advice goes beyond recommending lower levels of saturated fats, which can be recommended for everyone, and recommends not eating too much fish and vegetable oil. This recommendation goes against common advice and is given because of the specific genotype identified.

The test for APOE variant concluded that the tested person belongs to the 'normal type'. The report went on to state that women benefit from hormone replacement substitute therapy particularly after menopause. The report on CYP450 gene variants concluded that 'one gene works slowly (CYP 1A1) and the other works fast (CYP1B1). This means environmental elevated concentrations of dangerous products from your body's own estrogen or estrogen-like substances are generated. Resulting in a raised risk of various kinds of cancer (for women: breast and uterus; for men: prostate)'. The report recommended eating vegetables from the cabbage family and soy or red clover. It also suggested 'regular preventive medical check-up in cancer (women: breast and uterus; men: prostate gland)'. This report was more specific than Sciona's insofar as it identified specific cancers (breast, uterus and prostate) and, furthermore, recommended a pharmaceutical (hormone replacement therapy) in one case and regular cancer monitoring in the other.

The Genetic Health's report also indicated a protective effect of the results of the IL6 gene variant, stating: '*Great*, your polymorphism is associated with an increased life expectancy and a lower risk for inflammatory diseases. Even though, control your weight! In this case, insulin resistance can result in diabetes II'. This result and recommendation was again somewhat more detailed in that it identifies a protective effect rather than no risk. In this respect, Genetic Health has adopted a strategy different from Sciona's. By reporting the test results for each gene variant it did not combine them into generic advice but reported risks to specific cancers, suggested avoiding certain substances, which are commonly considered particularly healthy, and also recommended medical practices (monitoring) and medications (HRT). Furthermore, it also indicated low risk or protective effect, which has raised concern that individuals will consider themselves to be immune to disease based on incomplete evidence.

By taking a more medical approach to its advice than Sciona, Genetic Health exposed itself to criticisms of giving individuals information with clinical implications based on still emerging science. At the same time, its advice was more specific and personalized than Sciona's. The way the two reports phrased their results was also different enough to warrant speculating that Sciona's report would be less likely to worry people while the report by Genetic Health may be more likely to be of concern.

DISCUSSION

The development of a new science and technology based product is a process involving scientific, technological, economic, political and ethical concerns and stakeholders. The web pages of the nutrigenetic testing companies illustrate how a new technology can be driven by more than the science behind it. In the case of nutrigenetic testing, ethical and legal concerns have been important to the way in which the tests have been developed. In this sense, nutrigenetic tests are a case of ethics shaping the technology, as observed by Hedgecoe and Martin (2003).

The ethical and regulatory concerns around nutrigenetic testing have focused on the marketing of the tests. For example, in 2006, the US Government Accountability Office (GAO) accused the testing companies for misleading the public by making 'medically unproven' predictions (GAO, 2006). The sale of such tests direct to the public is currently often not regulated. In the UK, attempts have been made to categorize at least some of these tests as serious or high impact and requiring stricter regulation, while it has been suggested that 'lifestyle' genetic tests could be regulated more liberally (Human Genetics Commission, 2007). While this public controversy and the associated regulatory efforts discussed in this chapter are specific to nutrigenetic tests, they also articulate wider contemporary uncertainties about where medicine ends and something else, such as lifestyle, wellness or food industry, begins. In this new situation, stakeholders and regulators have sought to make the rules of the game clearer not only in relation to genetic testing but also in relation to complementary therapies (Clarke et al., 2004).

Based on the analysis of company web pages selling nutrigenetic tests, it becomes clear that some companies seek to appear as based in science and medicine and, at the same time, provide advice that is more consistent with wellness or lifestyle enhancement than it is with medicine. Companies market and develop nutrigenetic tests for enhancing general wellness or vitality and less as helping to avert a specific illness with specific means. The language used on these web pages is usually upbeat and reassuring and provides general recommendations about healthy lifestyle, such as the importance of eating vegetables, avoiding saturated fats, taking multivitamins and exercise. At the same time, the companies include images to suggest the advice is scientific, personalized and precise (aided by the use of images of scientists, graphs and references to published works and laboratory accreditation). On one hand, this manner of developing and marketing nutrigenetic tests raises questions about the added value of genetics to general lifestyle advice, particularly if the general advice is couched in language that attempts to convey a basis in science, or specificity to the individual tested. On the other hand, while the advice may not be particularly personalized, it may not be particularly harmful since it does not suggest individuals take unusual or dramatic actions. Furthermore, its general positive tone is unlikely to cause anxiety.

Other companies seek to appear to be more in line with mainstream medicine. They use clinical language and often address their potential customers as physicians or patients. They also more directly talk about diseases and risks. The advice provided is more specific than general advice on healthy eating, for example in the case of the report by Genetic Health. Genetic Health gave genotype specific recommendations, such as avoiding polyunsaturated fats ('the Mediterranean diet'), which is generally considered healthy. Some of the recommendations also referred to practices and products generally considered medical, such as use of drugs or cancer monitoring, rather than part of the wellness or food industries. On one hand, this manner of developing and marketing nutrigenetic tests seems better to live up to the promise of offering added value in comparison to off-the-shelf health advice if the recommendations are more specific for the individual genotype. The advice may also be more potent both in physiological and psychological terms. On the other hand, providing such specific advice based on emergent science raises the question of whether the evidence is sufficient to recommend people to be concerned about eating fish or to contemplate HRT and seek cancer monitoring. Thus, while this manner of developing nutrigenetic testing may be more specific, it has the potential to be more harmful.

There is an emerging academic and policy literature on how to regulate the practice and marketing of nutrigenetic testing (see Chapters 4, 5 and 6456). Katz (2007) has suggested that, contrary to common belief, regulation benefits companies, such as pharmaceutical companies. Katz makes reference to the late 19th century UK and USA drug markets, which were saturated with largely ineffective products that could often worsen the conditions they purported to cure. He argues that the regulation of the marketplace in turn-of-the-century America coincided with the transformation of the quack medicine market into the current, highly scientific one. Katz concludes that regulation is a tool that prevents a market for substandard goods. Without a regulated marketplace, firms lack an incentive to develop better products, as there is no mechanism to enable them to distinguish themselves from quackery.

Applied to nutrigenetic testing, Katz's argument suggests that stricter regulation could weed out 'snake oil' sellers from the nutrigenetics market and create incentives to develop tests with proven scientific benefit. The trouble is that the current market for nutrigenetic testing is reminiscent of the 19th century pre-antibiotic market for ambiguous cures, in that nutrigenetic tests, by and large, are not based on solid and tested scientific evidence. There are genetic tests of varying degrees of validity available but dietary recommendations based on individual genotype are currently considered premature (Haga et al., 2003). One might conclude that premature nutrigenetic tests might need to be regulated out of the marketplace. Regulators have not, however, pursued this approach (Human Genetics Commission, 2007). It raises the question of whether the eradication of strictly unproven but broadly harmless assessments and treatments would apply only to genetic testing (and on what grounds), or to any product or service available in the marketplace that claims to promote health and wellness. Academics (e.g. Ries, 2008) and policy-makers (e.g. Human Genetics Commission, 2007) have been arguing for regulating genetic tests based on the level of risk they pose. For example, in its recent report, UK Human Genetics Commission (HGC) suggested that genetic tests for 'medical' and 'lifestyle' purposes may require different forms of regulation. At the same time, the HGC acknowledges the difficulty in drawing this line, for example, deciding whether a test identifying an 'increased risk of cardiovascular disease' would be medical or not (Human Genetics Commission, 2007).

The websites illustrate how the companies adapt in a fluid situation where science, regulatory frameworks and ethical debates all are evolving.

Companies tend to oscillate between claims of enhancing wellness and preventing illness. This chapter does not seek to say that one model is necessarily better than the other but discusses some of the advantages and disadvantages of both approaches. Perhaps the most important aspect of the debates around nutrigenetic tests is how this process affects the way in which the health-focused applications of genetics and genomics are imagined and developed beyond a few online testing companies. A broader discussion of those possible effects is beyond the scope of this chapter. Nevertheless, the way in which the development of nutrigenetic testing proceeds in tandem with public and ethical debates highlights the need for scholars engaged in these discussions to reflect on how their commentary may shape the way in which genetics might be integrated into various health services to boost general health advice or individualized interventions.

ACKNOWLEDGMENT

The research for this chapter was funded by the UK Wellcome Trust, Biomedical Ethics Programme (grant no: 080126/Z/06/Z), which is gratefully acknowledged. I am also grateful to Matt Reed for research assistance. Views expressed are the author's.

References

Brown, N. and Michael, M. (2003). A sociology of expectations: retrospecting prospects and prospecting retrospects. *Technol Anal Strategic Manage* 15:1–18.

Clarke, D.B., Doel, M. and Segrott, J. (2004). No alternative? The regulation and professionalisation of complementary and alternative medicine in the Untied Kingdom. *Hlth Place* 10:329–38.

Consumer Reports (2006). When DNA means 'do not attempt'. *Consumer Rep* 71:6.

Ernst, E. (2006). Complementary and alternative medicine: examining the evidence. *Commun Practit* 79:333–36.

Government Accountability Office (2006). Nutrigenetic testing: tests purchased from four web sites mislead consumers. Retrieved January 15, 2008, from: http://www.gao.gov/new.items/d06977t.pdf.

Glaser, B. (1965). The constant comparative method of qualitative analysis. *Soc Prob* 12:436–45.

Haga, S., Khoury, M. and Burke, W. (2003). Genomic profiling to promote a healthy lifestyle: not ready for prime time. *Nat Genet* 34:347–50.

Hedgecoe, A. (2004). *The politics of personalised medicine: pharmacogenetics in the clinic*. Cambridge University Press, Cambridge.

Hedgecoe, A. and Martin, P. (2003). The drugs don't work: expectations and the shaping of pharmacogenetics. *Soc Stud Sci* 33:327–64.

Hogarth, S., Liddell, K., Ling, T., Sanderson, S., Zimmern, R. and Melzer, D. (2007). Closing the gaps: enhancing the regulation of genetic tests using responsive regulation. *Food Drug Law J* 62:831–48.

Human Genetics Commission (2002). Inside information: Balancing interests in the use of personal genetic data. A Report.

Human Genetics Commission (2007). More genes direct: a report on developments in the availability, marketing and regulation of genetic tests supplied directly to the public. A Report. London.

Katz, A. (2007). Pharmaceutical lemons: Innovation and regulation in the drug industry. Research Paper No 964664. University of Toronto Legal Studies Series.

Kramer, P. (1993). *Listening to Prozac: a psychiatrist explores antidepressant drugs and the remaking of the self*. Penguin, New York.

Lippman, A. (1998). *The politics of health: geneticization versus health promotion. In The politics of women's health: Exploring agency and autonomy*. Temple University Press, Philadelphia.

Lury, C. (2000). The united colors of diversity: essential and inessential culture. In S. Franklin, C. Lury & J. Stacey. Global nature, global culture. London: Sage.

Ordovas, J.M. (2004). The quest for cardiovascular health in the genomic era: nutrigenetics and plasma lipoproteins. *Proc Nutr Soc* 63:145–52.

Rajan, K.S. (2006). *Biocapital – the constitution of postgenomic life*. Duke University Press, Durham.

Ries, N. (2008). Regulating nutrigenetic tests: an international comparative analysis. *Hlth Law Rev* 16:9–20.

Rose, N. (2007). *The politics of life itself*. Princeton University Press, Princeton.

Secretary's Advisory Committee on Genetic Testing (2001). Development of a classification methodology for genetic tests. National Institutes of Health, Bethesda.

CHAPTER 12

Framing Nutrigenomics for Individual and Public Health: Public Representations of an Emerging Field

Timothy Caulfield, Jacob Shelley, Tania Bubela and Leia Minaker

OUTLINE

Nutrition and Genomics
ISBN: 978-0-12-374125-7

SUMMARY

Much is expected from the emerging field of nutrigenomics, particularly for the public's health. Representation of nutrigenomics by dominant stakeholders has important implications for public perceptions, uptake and, ultimately, viability of this new science. This chapter examines how the various stakeholders in the field represent the new science in the public sphere. Examining claims made in scientific literature, by research groups, private companies and the popular press, we address concerns that nutrigenomics is being represented in an overly optimistic light without sufficient regard to limitations. Specifically, we critique claims that nutrigenomic information will result in meaningful changes in dietary behavior, note the insufficient evidence that genetic information influences behavior generally and question the viability of nutrigenomics as a public health strategy. Recognizing the potential nutrigenomics may harness, we stress the need to address existing gaps in the research.

INTRODUCTION

The fields of preventive medicine and public health were thought forever altered with the completion of the Human Genome Project. Preventive medicine and public health, which traditionally focused on modifying risk factors, interventions and screening, could finally tackle genetic issues previously viewed as intransigent, immutable and innate (Coughlin, 1999). The intersection of genetics, public health and preventive medicine was hailed as bringing about an 'emerging paradigm of disease prevention – the identification and modification of environmental risk factors among persons susceptible to disease due to genotype' (Coughlin, 1999). The integration of genomics into the field of public health has been described as one of the most important challenges for the future of health care (Brand et al., 2006).

Nutrigenomics, considered one of the more promising applications of genomics, has been called the 'next frontier in the postgenomic era' (Kaput and Rodriguez, 2004; see also, Subbiah, 2007). As Peregin notes (2001, original emphasis), nutrigenomics is also at the leading edge of nutrition science: '[i]f you were to sum up the future of nutritional science into a single word, chances are it would be *nutrigenomics*'.

Over the past few years, the emerging area of nutrigenomics has received a considerable amount of attention, both from the media and the scientific community. As with other areas of genomics and genetics, nutrigenomics has raised ethical and social concerns (Chadwick, 2004; Ries and Caulfield, 2006; Reilly and DeBusk, 2007; Ozdemir and Godard, 2007). Indeed, some nutrigenomic services currently available to consumers have been critiqued as unnecessary and misleading (Government Accountability Office, 2006; BBC News, 2007). The Minnesota Department of Health (2006), for example, contends that while nutrigenomics holds great promise, 'since we know so little about how diet affects gene function, use of this information may be premature'. There has been some speculation as to whether nutrigenomics will represent the next manifestation of 'genohype' or will represent 'genohealth' (Ries and Caulfield, 2006). Some commentators maintain that the benefits of personalized nutrition, available through nutrigenomics, will not only improve health and aid in disease prevention, it may result in the 'amelioration of many of the "lifestyle" diseases' (Ghosh et al., 2007). Nutrigenomics is thus often framed as part of preventive public health measures. Although skepticism is often expressed about the immediate benefits of nutrigenomics, particularly about overly optimistic claims of effective therapies in the near future, there nevertheless remains a belief that nutrigenomics will have a positive impact both for individual and public health.

What are the claims to support the rise of the field and how might the representations of nutrigenomics shape its future? As Nisbet and Mooney (2007) articulate, public framing of a subject, 'allows citizens to rapidly identify why an issue matters, who might be responsible, and what should be done'. This chapter explores how this emerging field is represented in the public sphere by examining how nutrigenomics is framed in the scientific literature, by research groups, private companies and the popular press. The representation of nutrigenomics by these dominant stakeholders will likely have important implications for public perceptions, uptake and, ultimately, viability of this new science. The chapter addresses concerns that nutrigenomics is being represented in an overly optimistic light – a common phenomenon with emerging fields of study – without sufficient regard to limitations (see Bubela and Caulfield, 2004; Caulfield, 2005).

INTEREST IN NUTRIGENOMICS: THE RESEARCH AGENDA

There has been a rapid increase in the interest in nutrigenomics as a research topic. It is an area that has been viewed as worthy of public funding, both as a topic of basic scientific inquiry and as a field with health care and commercialization possibilities. From around 2004 forward, we see an upsurge of nutrigenomic programs, conferences and funding opportunities. For example, in 2004 the European Nutrigenomics Organization (NuGo) was established. Among other things, this multinational entity aims to train scientists to 'use post-genomic technologies in nutrition research' and to 'develop and integrate genomic technologies for the benefit of European nutritional science' (NuGO, 2004a). In the USA, there are a number of established university nutrigenomics initiatives, such as the NCMHD Center of Excellence for Nutritional Genomics at the University of California Davis and the Cornell Institute for Nutritional Genomics (2001). The US Department of Agriculture also has several nutrigenomic programs under its Agricultural Research Service division (e.g. Tufts University's Jean Mayer USDA Human Nutrition Research Centre on Aging) and the US National Institutes of Health has funded a variety of nutrigenomics projects, often through the Nutritional Science Research Group.

Many other countries have followed suit. For instance, New Zealand has a nutrigenomics organization, Nutrigenomics New Zealand. Similarly, in Canada, the Advanced Foods and Materials Network (AFMNet), a federally funded Network of Centers of Excellence (NCE) program, also has a number of nutrigenomics projects. Several Canadian universities have academic research chairs and programs in nutrigenomics, including the Universities of Toronto, Manitoba and Alberta and Memorial University.

The private sector is also increasingly involved in nutrigenomics through research, development and commercialization activities in genetic testing services, nutritional supplements and novel food products. To cite just one example, Genelex Corporation offers individualized nutritional and lifestyle counsel that 'lasts a lifetime because your genes are not a fad' (Genelex, 2006). Specifically, the company suggests that:

> Testing examines your personal variations in nineteen genes that scientists have shown play major roles in your body's heart and bone health, detoxification and antioxidant capacity, insulin sensitivity, and tissue repair. Your DNA test results, combined with information from your completed lifestyle questionnaire, result in personalized, realistic steps you can take to improve and maintain your good health (Genelex, 2006).

In 2000, Dave Evans, CEO of WellGen, claimed, '[i]n less than 10 years, you'll be able to go to a lab and complete a set of genetic tests to identify your personal disease susceptibilities. When you leave you'll be armed with a list of foods to eat and foods to avoid and a recommendation of dietary supplements to help prevent your diseases' (Fogg-Johnston and Merolli, 2000).

THEMES IN NUTRIGENOMICS

While the above list of funding opportunities, nutrigenomics research groups and private companies is far from comprehensive, it provides a sense of the rapid growth of interest in the area and the degree to which the value of this emerging area has been accepted by various stakeholders. Indeed, given the amount of investment – in time, money and resources – there seems to be a clear expectation that nutrigenomics will bear fruit, from both a health and economic perspective. Some anticipate that nutrigenomics will have an economic benefit, especially for those companies that are initially successful in penetrating the nutrigenomics niche market. Although the science of nutrigenomics might evolve rapidly from its current status, 'the business outlook is viewed as encouraging' (Brown and van der Ouderaa, 2007). Moreover, it is thought that nutrigenomics will be 'a cost-effective approach to chronic disease, and one that can reduce the nation's healthcare cost' (Johnson et al., 2006).

This rise in interest presents an opportunity to explore how a new area of research is framed. In other words, what is the 'public face' of this new field? What are the stated benefits and limitations? Finally, what are the potential consequences of these representations of nutrigenomics? What themes have emerged?

Scientific Literature

We collected 35 review and commentary articles from peer-reviewed scientific journals discussing the public health impact of nutrigenomics. Of these, 27 articles were collected through keyword searches of major databases of academic and scientific scholarship, including the Embase, Medline and EBSCO databases. The following key terms were used: 'nutrigenomics', 'nutritional genomics', 'nutrigenetics', 'public health', 'public policy', 'regulation' and 'health policy'. We narrowed the search by only examining those articles that were either review or commentary articles. The remaining eight articles were identified by members of our interdisciplinary research team. Articles were reviewed to identify specific claims made regarding the possible impact of nutrigenomics on public health. The claims we identified were later categorized into general themes, which were determined from the content of the claims.

In general, we found that the main themes within the scientific journals were that nutrigenomics would lead to: improved dietary advice; the development of health promoting supplements; preventative health strategies; and the reduction of health care costs. This last theme is particularly interesting because it is a common justification for supporting new technologies – likely because the idea of paying now to save in the future is a politically palatable rationale. This is especially so for publicly funded health care systems, where cost-effectiveness is a primary policy issue. In the area of nutrigenomics, it is suggested that cost savings will come from the reduction of chronic disease. For example, Johnson et al. (2006) suggest that: '[p]erhaps an overriding benefit still not adequately known in the general population is that this information can be used to prevent disease. Prevention is a cost-effective approach to chronic disease, and one that can reduce the nation's healthcare cost'. It is claims such as these that are the main focus of our examination.

In spite of the fact that nutrigenomics will be largely focused on individual dietary advice, some academics in review articles represent it as a boon to public health. More than simply managing or treating disease or the symptoms associated with disease (i.e. post-diagnosis), nutrigenomics will be used to identify susceptibilities to diseases and implement proactive measures to help individuals avoid developing the disease in the first place (DeBusk et al., 2005). Although it is recognized that the uptake of nutrigenomics as a public health strategy may not be immediate, there is generally an expectation that the field will make an important contribution in the long term (Müller and Kersten, 2003). As Ordovas and Mooser (2004) contend, nutrigenomics will have public health implications because of its 'potential to change dietary habits in order to achieve effective disease prevention and therapy'. Others have a less tempered expectation, claiming that, '[u]ltimately, nutrigenomics research will lead to development of evidence-based healthful food and lifestyle advice and dietary interventions for contemporary humans' (Afman and Müller, 2006).

Already noted above is the claim by Ghosh et al. (2007) that nutrigenomics will help ameliorate 'lifestyle' diseases. Kaput et al. (2007) contend, '[d]iagnostics, preventive lifestyle guidelines, more efficacious dietary recommendations, health-promoting food supplements, and drugs are some of the anticipated end-products of nutrigenomics research'. The efficacy of nutrigenomics dietary recommendations is perceived to be so strong that Arab (2004) projects that parents will test infants at birth for nutrigenetic profiles 'in order to intervene before the fetal origins of disease can develop into the later decade realities and so that dietary damage cannot accumulate'.

Kaput (2005) claims that nutrigenomic information not only harnesses the potential to optimize health and prevent or mitigate chronic diseases, but that, '[o]ptimal nutrition may also influence the aging process'. Healthier aging, made possible with appropriate nutrition, is also a benefit highlighted by Ordovas and Mooser (2004). Although they recognize there are limitations, they nevertheless conclude that the 'preliminary evidence strongly suggests that the concept of nutritional genomics should work' (Ordovas and Mooser, 2004). This optimism, even when limitations are identified, is widespread in the literature surveyed. Hence, the observation by Brand et al. (2006) that, 'the integration of genomics into public health research, policy and practice will be one of the most important challenges for our health care systems in the future' seems particularly relevant for nutrigenomics.

Research Groups

As noted above, many new research groups have a specific focus on nutrigenomics. These groups range from federally funded partnerships with industry (e.g. Nutrigenomics New Zealand) to academic collaborative ventures (e.g. UC Davis NCMHD Center of Excellence for Nutritional Genomics) to less formal collections of interdisciplinary teams (e.g. Memorial University's Nutrigenomics Research Interest Group). The research groups reviewed were identified using Internet based search engines (e.g. Google), supplemented by reference to the various affiliated groups of the authors in the scientific literature identified above. We utilized a purposive sampling strategy, examining the following groups: Nutrigenomics New Zealand, BioProfile Nutrigenomics, UC Davis NCHMD Center of Excellence for Nutritional Genomics and NuGO. We relied on the groups' websites to assess how they framed nutrigenomics. Again, we recognize that this review is not comprehensive, but feel that it is a representative sample of the most immediately available research initiatives.

The public face of these groups, usually on websites, is optimistic about the social benefits that may accrue from the research. Nutrigenomics New Zealand (2004a) believes that 'nutrigenomics will lead to the development of new foods for individualized health and nutritional benefit'. NuGO states that nutrigenomics 'promises to improve health conditions and to prevent disease, e.g. diabetes, obesity, cardiovascular diseases and cancer' (NuGO, 2004b). For a number of the groups, commercialization is also a strong theme. For example, Germany's BioProfile Nutrigenomics (2005) holds that: '[n]utrigenomics is a highly innovative and fast-growing interdisciplinary field of research linking genome research, plant biotechnology and molecular nutritional research and offering new applications for medicine and nutrition'. As an entity, BioProfile strives to 'expand the industrial application of academic knowledge in the life sciences' (BioProfile, 2005). Not all the research groups had a strong commercialization mandate, however. The stated mission of the UC Davis NCMHD Center of Excellence for Nutritional Genomics, for example, is to 'reduce and ultimately eliminate racial and ethnic health disparities' (UC Davis, 2007a).

Research groups must seek to position nutrigenomics within a nation's research agenda and compete for public and private research funds. As such, the positive messaging about anticipated benefits is to be expected. As with the scientific literature, additional themes included the reduction in chronic disease and long-term cost savings for the health care system. The UC Davis NCMHD Center of Excellence for Nutritional Genomics, for example, recognizes that nutrigenomics will impact society, having applications that will likely exceed those of the Human Genome Project. 'Chronic diseases (and some types of cancer) may be preventable, or at least delayed, by balanced, sensible diets. Knowledge gained from comparing diet/gene interactions in different populations may provide information needed to address the larger problem of global malnutrition and disease' (UC Davis, 2007b). Nutrigenomics New Zealand (2004b) claims that the information garnered through nutrigenomics, 'will ultimately lead to the development of completely new, added-value, export-focused, gene-specific foods that will deliver proven health outcomes to consumers'.

Private Companies

The number of private companies dealing with genomics has increased dramatically in recent years. The UK Human Genetics Commission report, *More Genes Direct* (2007), identifies 26 companies advertising genetic tests to the public. Six of these companies were explicitly identified in the report as offering nutrigenomic testing (Genetic Health, Genelex, Holistic Heal, Quixtar, Saluge and Suracell), although a greater number of these companies are likely offering nutrigenomic testing or are in the process of developing such tests. Given the ever-changing landscape in companies offering nutrigenomic products, we reviewed the six companies identified in *More Genes Direct*. In addition to examining the websites of the six companies identified above, we examined two additional companies that offer nutrigenomic testing: Genovations and Sciona.

The eight companies we reviewed offer private testing services that include the provision of nutrigenomic tests and advice. Genovations (2002a) states that it focuses on testing for genetic variations 'influenced by environmental factors' and provides profiles that contain 'intervention options based on the patient's genomic pattern'. The company also claims to offer 'specific risk reduction strategies, including dietary, nutritional, lifestyle, and pharmaceutical interventions' (Genovations, 2002a). Other companies such as Sciona Inc., through its Mycellf program, are more focused on nutritional and lifestyle advice. Sciona Inc. (2007c) maintains that it 'provides personalized health and nutrition recommendations based on an individual's diet, lifestyle and unique genetic profile'. Some companies, such as Holistic Heal (2008), emphasize that the information their tests provide 'is not intended to diagnose, treat, cure or prevent disease'. Suracell (2007) and Salugen (2006) included a similar statement in fine print. All three companies, however, allege that their products will help individuals manage their health and well-being.

Naturally, the companies emphasize the potential for disease prevention. Genovations (2002b) contends that its testing services will 'empower physicians and patients to realize ... more effective preventive interventions ... [and] ... improved clinical insight into patients with treatment-resistant "chronic" conditions'. Genelex (2006) contends its recommendations are 'practical, effective and proven ways in which to improve your short and long-term health, help prevent illness, and above all feel better'. Sciona's Mycellf program contends that 'your unique genetic profile is the key to understanding how your body works, including which diet and exercise programs will bring you the results you want and which health and nutrition programs will lead to long-term wellness' (Sciona, 2007b).

The representations associated with the private companies differ from the scientific literature and research groups in that they emphasize individual benefits rather than social benefits such as the reduction of the chronic disease burden. As companies that currently focus on the direct-to-consumer market, this is unsurprising. Recall the claim by Genelex (2006) that the advice they offer is not based on a dietary fad, but instead on '*your* body's real needs' (Genelex, 2006, emphasis added). Salugen (2006) attempts to distinguish itself from other companies that offer a 'one-formulation-fits-all approach to wellness' by providing what they term individualized medicine, 'a one-formulation-fits-one approach .... [that] delivers precisely what you need to improve your wellness or alleviate your condition'. The promise of Holistic Heal (2008) is a 'personalized map' to help achieve wellness.

Popular Press

Nutrigenomics has received an increasing amount of attention in the popular press since 2004. We searched Lexis/Nexis, Factiva and Canadian Newsstand media databases for all media coverage with no date restrictions using the following search string: "nutrigenomic* or nutragenomic* or 'nutritional genomic*' or 'personalized nutrition' or nutragenetic or nutrigenetic or 'gene food"'. We then hand-sorted the media coverage into broad categories and eliminated false hits that were not related to nutrigenomics. In addition, we compiled a list of 14 companies through a non-random sampling method using the above search terms coupled with 'gene* and test' and company. We used snowball sampling to augment our list by extracting references to nutrigenomics companies from media articles, policy reports and academic articles. We then repeated our media searches in Lexis/Nexis, Factiva and Canadian Newsstand to search for media articles on nutrigenomics companies with the search string "'Company Name' and 'gene* and test"'. Again, these articles were assessed for relevance. For an explanation of the methodology employed to analyze systematically media representation in this context see Bubela and Taylor (2008). Here the framing shifts from a perspective on public health benefits to one focused on the individual.

> Media coverage plays into the needs of consumers, especially those who are wealthy, educated and interested in being trend-setters. It feeds on fears of disease and aging in, probably, the most health-conscious demographic, and nurtures notions of individuality in an era of eroding public health care systems (Bubela and Taylor, 2008).

Unlike some research areas, such as medical genetics more generally, most of the press coverage has not been about specific research 'breakthroughs', although there are a few high profile exceptions, such as media reports on the association between coffee intake, the CYP1A2 genotype and the risk of suffering myocardial infarction (see Cornelis et al., 2006). Much newspaper coverage of nutrigenomics has been in the form of feature articles about the emerging field specifically, or genetic testing more generally.

As with the other forms of public representations noted above, the media coverage has been largely optimistic. Media representations of nutrigenomics emphasize benefits, often framing nutrigenomics 'as providing legitimate results which may be relied on for diet and overall health information' (Bubela and Taylor, 2008). This positive frame and tendency to exaggerate benefits may be 'because the main sources of information are nutrigenomics companies and, in some cases, the entrepreneurial scientists who founded them' (Bubela and Taylor, 2008). Indeed, one third of newspaper articles in a study by Bubela and Taylor (2008) referred to nutrigenomics companies while less than 7% referenced published research.

Adding to the overall positive framing of nutrigenomics in the media were stories focused on celebrity lifestyles and diets, identifying nutrigenomics as a new diet trend. In such stories, nutrigenomics is framed as a field that will revolutionize and personalize dieting. This framing appeals to wealthy and sophisticated consumers who desire 'ownership . . . the feeling that something is tailored to their needs and will overcome their problems' (Dickins, 2006).

But, because the media reports are often reviews or feature articles, the risks or limitations of the field have also received some attention. For example, the *Globe and Mail*, a Canadian national newspaper, published a feature article entitled 'Personal Genetics Tests: Genius or Bogus?' that questioned the value of the emerging direct-to-consumer genetic testing industry (Mick, 2007). Some commentators suggest the science of nutrigenomics is too immature to support tailored dietary prescriptions. Jim Kaput, president and chief scientific officer of NutraGenomics Inc., a fledgling nutrigenomics research company in Chicago that does not offer testing services to the public, says, '[p]eople will spend money for this, but in terms of science-based nutritional advice it's just too early' (MacGregor, 2005). Kaput adds that the tests are 'for rich people with an extra $1000 who want to say, "I did my genotype"' (MacGregor, 2005).

Likewise, specific controversies, such as the 2006 nutrigenomics investigative report by the US Government Accountability Office (GAO), also received media attention. The investigation of four companies operating in the USA concluded that '[t]he results from all the tests GAO purchased mislead consumers by making predictions that are medically unproven and so ambiguous that they do not provide meaningful information to consumers' (GAO, 2006). The companies recommended the use of costly nutritional supplements where cheap alternatives exist, provided generic advice based more on the lifestyle profile than on the DNA and indicated susceptibility to a variety of diseases, albeit with disclaimers that the tests were not intended to diagnose disease.

This type of polarized story telling, i.e. an emphasis on either potential breakthroughs or social controversy, is typical of health care reporting (Highfield, 2000). As Bubela and Caulfield (2004) found, there is a tendency in both scientific and newspaper articles to overemphasize the benefits and under-represent the risks of genetic research. The popular press needs to present the material in a newsworthy manner. But, despite this tendency and the presence of some negative representations, the general themes found in the popular press were positive and emphasized health benefits to the individual (Cronshaw, 2006; Ursell, 2006).

CRITIQUING THE CLAIMS

In total, the major sources of the public representations – the science literature, the messaging from research groups, the marketing from testing companies and media stories – have framed nutrigenomics as an emerging field that will lead to individual and public health improvements, primarily through a reduction of chronic disease, an increase in healthy lifestyle choices and a lowering of health care cost. The tone is largely positive. But, as noted above, not all in the scientific community or in the popular press have been uniformly enthusiastic about nutrigenomics. Even some of the science papers noted above contain statements about the limits. For example, Joost et al. (2007) observe, '[t]he possibility cannot be excluded that, in some individuals, knowledge of a genetic predisposition might lead to a fatalistic attitude and a reduced compliance with any intervention'.

The value of nutrigenomics as a research agenda is recognized. It is an excellent window into understanding the complexity of the human genome. The largely positive representations of nutrigenomics in material we reviewed may, however, lead to a misinterpretation of the potential limitations of the field and create inappropriate expectations. Given that nutrigenomics is being justified on the basis of the reduction of chronic diseases, our critiques here focus on this level of representation. They include: the insufficient consideration that has been given to the difficulties in changing behaviors, particularly dietary behaviors, the hyped expectation for genetic information to bring about meaningful change and the limited role of nutrigenomics as a public health initiative.

Dietary Information and Behavioral Change

Part of the skepticism towards the way nutrigenomics has been represented stems from the realization that even if nutrigenomic information provided relevant dietary information – and there are those who think we are a long way from having meaningful data since current attempts to derive dietary recommendations based on the genotypes of the few single nucleotide polymorphisms presently known to be associated with particular complex diseases appear largely experimental. Recently, such tests have, provocatively, been called 'genetic horoscopes' (Russo, 2006) – there exists doubt that this information would be more significant than what we already know about healthy diets. In other words, for nutrigenomics to have a broad positive impact it would need to provide dietary information that is more meaningful than information that is already available.

Most of the proposed benefits of nutrigenomics require some degree of behavior change by individuals. In large part, nutrigenomics is about the provision of information regarding an individual's predispositions in the hope that they will act on that information. The reduction in chronic disease and the lowering of health care costs, two of the dominant themes in the framing of nutrigenomics, require individuals to act on nutrigenomic informed advice – be it individual testing or the provision of risk information for a specific, identifiable, sub-population. The need to induce behavior change has been noted in some of literature. Indeed, it has been suggested that genetic information might help facilitate change. As Ordovas and Mooser (2004) note, nutrigenomics 'will certainly have public health implications because they have the potential to change dietary habits in order to achieve effective disease prevention and therapy'. Consider the implications of the representation by Genelex (2006) that '[g]enetic testing combined with a lifestyle assessment, provide you with a scientifically based, personal blueprint for optimizing health'.

Given this reality, it is important to consider what available evidence tells us about behavior change, health and food choices. Will people really change their eating habits in response to nutrigenomic information? Given that answering this basic question is fundamental to the stated goals of nutrigenomics, one would expect it to be a significant part of the nutrigenomic research agenda and a commonly noted limitation. In fact, only a minimal discussion of this issue was found in the above noted documents. Kaput et al. (2005) observe, '[s]uch knowledge is necessary, but not sufficient, to address health disparities among all racial and ethnic populations throughout the world. Social, economic and cultural factors are critical in selecting foods and designing studies to identify causative genes and interacting environmental factors'. Nevertheless, the limited function of this information is often minimized or, as is the case on company websites, included in a disclaimer. For example, Genelex's website (2006) includes the following disclaimer:

> The information presented on this site is intended as general health information and as an educational tool. It is not intended as medical advice. Only a physician, pharmacist, or other healthcare professional should advise a patient on medical issues and should do so using a medical history and other factors identified and documented as part of the health professional/patient relationship.

If anything, it is assumed that nutrigenomic information, given its genetic foundation, will be more likely to motivate change (this will be discussed in more detail below).

Considerable research has shown, however, that strategies aimed at improving health behaviors generally have limited success. Health behaviors occur within a social, economic and physical context and are thus difficult to change. Lifestyle choices are heavily influenced by life circumstances. This holds true for dietary behaviors. A multitude of factors have been explored to explain why people choose the foods they do. In a large, population-based study, taste and cost were found to be the most important determinants of food choice, with convenience, nutrition and weight control also being significant determinants, although determinants varied significantly by demographic and health lifestyle differences (Glanz et al., 1998). Broader influences on food choice have also been examined, including socioeconomic status (Drewnowski, 2004; Drewnowski and Darmon, 2005a), food marketing (French et al., 2001; Story et al., 2002), overarching societal influences (Booth et al., 2001), changes in food production, distribution, processing (Nestle, 2000) and the food industry (Lobstein, 1998; Alston et al., 2006). A further barrier is food preference due to cultural, religious and symbolic meaning. Food 'enables persons to represent themselves and express (aspects) of their lifestyle and identity' (Meijboom et al., 2003).

Two significant determinants on food choice are food availability and food affordability. In general, greater availability of food results in increased consumption. It is widely recognized that the current North American environment provides frequent opportunities for the consumption of large quantities of food (Hill and Peters, 1998; Cullen et al., 2000; Kubik et al., 2003). This is particularly true for lower income households. As Reidpath et al. (2002) summarize, '[t]he social determinants (SES) [socioeconomic status] and environmental determinants (density of fast-food outlets) interact to create environments in which the poor have increased exposure to energy-dense foods'. Population research on food cost, energy cost and pricing intervention research has confirmed the impact of food affordability on food choice. At a population level, there is an inverse relationship between the energy density of foods (kilocalories per gram) and energy cost (dollars per kilocalorie), resulting in the fact that diets high in refined grains and added fats and sugars are more affordable than the recommended diets based on whole grains, fresh vegetables and fruits and lean meats and dairy (Drewnowski, 2004). Simply put, healthy diets cost more than unhealthy diets (Drewnowski and Levine, 2003; Darmon et al., 2004; Drewnowski and Darmon, 2005a, 2005b; Drewnowski and Rolls, 2005).

The difficulty associated with changing health behaviors, therefore, is no different with respect to nutrigenomics, which is only as beneficial as its impact on food choice. In other words, if the information about how to 'eat for your genotype' is not put into practice, nutrigenomics fails as a public health measure and as a tool for individual behavior change. There is also a danger of 'victim-blaming,' the overlooking of personal 'response-ability' – the ability of an individual to respond to health messages appropriately given sufficient resources and support – in favor of personal responsibility to improve health behaviors (i.e. eating more fruits and vegetables, quitting or reducing smoking or engaging in more physical activity) (Labonte and Penfold, 1981). It is clear that food choice depends upon many factors and for any successful population-level or individual intervention to change food behaviors these factors need to be considered. This is especially the case for populations with a lower socioeconomic status, who are at greater risk to suffer from poor health and are more likely to struggle with food insecurity.

Nutrigenomic informed dietary approaches may not adequately take the determinants of food choice into account. Instead, it focuses on individual knowledge. Past food choice interventions focusing on increasing participants' knowledge that have neglected other determinants of food choice have been largely ineffective. Some argue, however, that dietary information tailored to one's genotype will prove more successful at creating long-term improvements in diet than previous individual and information-based strategies. Whether this will prove to be true remains to be seen.

Genetic Information and Behavioral Change

The expectation that nutrigenomics will result in meaningful dietary change rests on the assumption that 'individuals might feel a stronger motivation to comply with dietary recommendations, if these are tailored to their own genetic make-up' (Meijboom et al., 2003). Is there reason to believe that genetically based dietary advice will be more efficacious, particularly given that nutritional advice, for the most part, has been consistent over many years? As one article wryly observes, '[s]ome people will be advised to eat broccoli, while others will be to eat ... even more broccoli' (Underwood and Adler, 2005 cited in Ries and Caulfield, 2006). Will genomic information be able to overcome the above noted barriers positively to influence dietary behaviors?

Some speculate that nutrigenomic information may motivate change. For instance, Laurance (2005) holds that:

> [i]t has always been known that success in dieting depends on the dieter's psychological determination to change. But the new research suggests that if a way could be found for individuals to select the right diet for them, it could ease the demands on willpower.

While there is little research on the impact of genetic information on diet change, we can reflect on the impact genetic information has on behavioral change more generally. Marteau and Lerman (2001) note that, similar to any information on risk, genetic risk information could increase or decrease one's motivation to change behavior, the response being shaped by pre-existing perceptions. Overall, however, they conclude that '[t]he current evidence suggests that providing people with DNA derived information about risks to their health does not increase motivation to change behavior beyond that achieved with non-genetic information' (Marteau and Lerman, 2001). Carlsten and Burke (2006) reached the same conclusion: '... genetic risk information may be ineffectual in motivating behavior change or potentially may even be harmful by inducing fatalism, feelings of impotency, or loss of willpower'. They caution against "genocentric views" which may lead to inappropriate expectations rather than substantive progress towards improving health outcomes' (Carlsten and Burke, 2006).

Not only is there a lack of evidence to suggest that individuals will be motivated to make lifestyle changes based on genetic information, there has been some concern that there might be an adverse flip side. Namely, 'that patients who test negative may be falsely reassured and thus less motivated to comply with preventive recommendations' (Hunter et al., 2008). A further concern is that even in light of susceptibility that genetic information may, in fact, reduce motivation (Hunter et al., 2008). This has been called 'fatalism' – the idea that people will come to view their genetic predispositions as signaling an inevitable course (Wright et al., 2003; Carlsten and Burke, 2006). The concept is expressed well by a Newfoundlander who discovered, as the result of research on the link between heart disease and genetic heritage, that he had a 'genetic flaw' predisposing him to heart disease: 'This means we are doomed so we might as well live it up. We don't need to quit smoking or change our diets' (Gillis, 1999). While the concept of fatalism is little discussed in the nutrigenomic literature, there is some recognition of this issue. For example, Joost et al. (2007) note: 'The possibility cannot be excluded that, in some individuals, knowledge of a genetic predisposition might lead to a fatalistic attitude and a reduced compliance with any intervention'.

With this general conclusion about the impact of genetics on behavior change as a backdrop, it is important to note that one study involving obese patients did find that nutrigenomic information was catalytic in maintaining weight loss (Arkadianos et al., 2007). The study, sponsored by the nutrigenomics company Sciona which involved patients who had been unsuccessful in previous weight loss attempts, found that the study group that received nutrigenomic information did 'significantly better' than the groups that were not given this information (Arkadianos et al., 2007). The authors conclude that 'the use of nutrigenetics to improve and optimize a healthy balanced diet in a clinical setting could be an effective aid in long-term lifestyle changes leading to sustained weight loss' (Arkadianos et al., 2007).

The results are, no doubt, promising and it is exactly the kind of research that is needed (Newell et al., 2007). However, given existing data about the impact of genetic information on behavior and the fact that this was a highly motivated group in a clinical setting, it is difficult to say whether such data are generalizable to the broader public, which relates more closely to the core claims of the nutrigenomic community. More research is needed to see if the study's findings will translate into a population health intervention. Also, it could be argued that success in weight loss, which is observable and quantifiable, differs from showing evidence of disease prevention, which is a perennial challenge for public health interventions. Additionally, genetic information does not eradicate the food intake barriers identified above. Even if an individual were compelled by nutrigenomic information to make dietary behavior changes, if they are unable to afford or access healthier foods, the initial ambition will be easily frustrated. Similarly, even if significant changes are implemented initially, sustaining them over a long period of time may be problematic.

Nutrigenomics as a Public Health Strategy

In the material we reviewed, nutrigenomics is often related to public health benefits. Public health has been defined as 'what we, as a society, do collectively to assure the conditions in which people can be healthy' (Institute of Medicine, 1988). Laverack (2004) notes that health may be considered from at least three perspectives: biomedical, behavioral and socioenvironmental. A biomedical perspective on health focuses on the 'normal physical state' – specifically, the absence of physical or mental disability, pain or disease, health being a state free of physiological or psychological deviation from the norm. In the case of nutrigenomics, for example, a biomedical perspective on health means identifying genetic predisposition to diet-related disease and intervening to mitigate risks. A behavioral definition of health understands health as energy, functional ability and a healthy lifestyle and emphasizes the impact of lifestyle choices (e.g. diet, smoking and physical activity habits) on health status. Finally, a socioenvironmental perspective on health takes into account the social and environmental causes of health, focusing on psychosocial risk factors and socioenvironmental risk conditions instead of on high-risk individuals. All three conceptions of health have a role in maintaining the public's health, given that they all help to assure conditions for people to be healthy. Interestingly, although nutrigenomics fits into the biomedical definition of health, the proposed outcome of nutrigenomics (i.e. improved diet due to individual counseling) represents a behavioral definition of health.

Although proponents of nutrigenomics anticipate the science will contribute to disease prevention, its success as a population health strategy will face challenges. The goals of population health 'are to maintain and improve the health status of the entire population and to reduce inequities in health status between population groups' (Health Canada, 2001). Inequities in health status, in part, are due to inequalities in the social determinants of health. Nutrigenomics may inadvertently exacerbate health disparities between the rich and poor by failing to take contextual determinants of behavior into account. Moreover, the individualistic and technological nature of nutrigenomics means that implementing a nutrigenomics program at a population level could be extremely expensive, unless a strategy focused on nutrigenomic-informed population advice. Even if it were demonstrated that nutrigenomic testing results in improved diets at an individual level, would health care funders be willing to add the cost of nutrigenomic testing to insured services? If nutrigenomic tests and counseling were to be sold directly to the public, it is likely that the vast majority of people who buy them will be those with higher income who are healthier than lower income people.

The emphasis that nutrigenomics constitutes a public health measure is no doubt a reflection of the growing recognition of the importance of public health initiatives. It is also strategic, given that much nutrigenomic research is funded through the public purse. Additionally, it pre-emptively rebuts criticisms that nutrigenomics is simply a first world intervention intended for wealthy individuals. As Chávez and de Chávez (2003) note, 'the most important problem of these biotechnological solutions is that they may only favour members of the more affluent social groups and countries'. This is a common criticism levied against nutrigenomics. As Ries and Caulfield (2006) note, media articles often refer to the appeal of nutrigenomics to 'anxious yuppies' and the 'very small and very health-conscious segment of the population'. Darnton-Hill et al. (2004) perhaps express the criticism best:

> The high costs of the screening and genotype diagnosis of developing novel and functional foods and the poor availability of functional health systems make even the possibility of 'tailored diets' an impossible dream for most populations relying on poorly functioning and poorly resourced health systems.

CONCLUSION

Nutrition is clearly a tremendously important part of public health. Gaining a more complete understanding of the relationship between genes, food and lifestyle in the development of disease is a worthwhile course. It is understandable that researchers are excited about the possibilities of the field to improve health. To date, however, there seems little compelling evidence to support many of the claims about long-term benefits commonly made by the various stakeholders. To be fair, nutrigenomics is in its infancy and it is certainly possible that nutrigenomic testing may help motivate individuals to improve their dietary habits. But the existing gaps in knowledge and the potential challenges to the realization of health benefits, such as the difficulty in motivating behavior change and overcoming barriers to food choice, deserve more attention.

References

Afman, L. and Müller, M. (2006). Nutrigenomics: from molecular nutrition to prevention of disease. *J Am Dietet Assoc* 106:569–76.

Alston, J.M., Sumner, D.A. and Vosti, S.A. (2006). Are agricultural policies making us fat? Likely links between agricultural policies and human nutrition and obesity, and their policy implications. *Rev Agricult Econ* 28:313–22.

Arab, L. (2004). Individual nutritional recommendations: do we have the measurements needed to assess risk and make dietary recommendations?. *Proc Nutr Soc* 63:167–72.

Arkadianos, I., Valdes, A.M., Marions, E., Florou, A., Gill, R.D. and Grimaldi, K.A. (2007). Improved weight management using genetic information to personalize a calorie controlled diet. *Nutr J* 6:6–29.

BBC News (2007). Firm offers online DNA analysis. BBC News, (Online). 11 November. Available at: http://news.bbc.co.uk/2/hi/science/nature/7098998.stm Accessed 11 November 2007.

BioProfile Nutrigenomics (2005). *Bioprofile Nutrigenomics. (Online).* Available at: http://www.nutrigenomik.de/media/downloads/downloads_1/Nutri_Flyer_2005-10.pdf Accessed 26 February 2008.

Booth, S.L., Sallis, J.F., Rittenbaugh, C., et al. (2001). Environmental and societal factors affect food choice and physical activity: rationale, influences, and leverage points. *Nutr Rev* 59:S21–S39.

Brand, A., Schröder, P., Brand, H. and Zimmerman, R. (2006). Getting ready for the future: integration of genomics into public health research, policy and practice in Europe and globally. *Commun Genet* 9:67–71.

Brown, L. and van der Ouderaa, F. (2007). Nutritional genomics: food industry applications from farm to fork. *Br J Nutr* 97:1027–35.

Bubela, T. and Caulfield, T. (2004). Do the print media hype genetic research? A comparison of newspaper and peer-reviewed research papers. *Can Med Assoc J* 170:1399–407.

Bubela, T. and Taylor, B. (2008). Nutrigenomics, mass media and commercialization pressures. *Hlth Law Rev* 16:41–9.

Carlsten, C. and Burke, W. (2006). Potential for genetics to promote public health: genetics research on smoking suggests caution about expectations. *J Am Med Assoc* 296:2480–2.

Caulfield, T. (2005). Popular media, biotechnology and the 'cycle of hype'. *J Hlth Law Pol* 5:213–33.

Chadwick, R. (2004). Nutrigenomics, individualism, and public health. *Proc Nutr Soc* 63:161–66.

Chávez, A. and de Chávez, M.M. (2003). Nutrigenomics in public health nutrition: short-term perspectives. *Eur J Clin Nutr* 57(Suppl. 1):S97–S99.

Cornelis, M.C., El-Sohemy, A., Kabagambe, E.K. and Campos, H. (2006). Coffee, CYP1A2 genotype, and risk of myocardial infection. *J Am Med Assoc* 295:1135–41.

Cornell Institute for Nutritional Genomics (CING) (2001) *Cornell Institute for Nutritional Genomics (CING). (Online).* Available at: http://www.research.cornell.edu/VPR/CenterDir/CING.html Accessed 26 February 2008.

Coughlin, S.S. (1999). The intersection of genetics, public health, and preventive medicine. *Am J Prevent Med* 16:89–90.

Cronshaw, T. (2006). Customized food 'hot'. *Press Newspaper Health Section* 14 Jul.

Cullen, K., Eagan, J., Baranowski, T., Owens, E. and de Moor, C. (2000). Effect of à la carte and snack bar foods at school on children's lunchtime intake of fruits and vegetables. *J Am Diet Assoc* 100:1482–6.

Darmon, N., Briend, A. and Drewnowski, A. (2004). Energy-dense diets are associated with lower diet costs: a community study of French adults. *Pub Hlth Nutr* 7:21–7.

Darnton-Hill, I., Margetts, B. and Deckelbaum, R. (2004). Public health nutrition and genetics: implications for nutrition policy and promotion. *Proc Nutr Soc* 63:173–85.

DeBusk, R.M., Fogarty, C.P., Ordovas, J.M. and Kornman, K.S. (2005). Nutritional genomics in practice: where do we begin? *J Am Diet Assoc* 105:589–98

Dickins, J. (2006). New clues to the perfect diet – what's good for us in the genes. *Sunday Telegraph*, 24 Sep. p. 26.

Drewnowski, A. (2004). Obesity and the food environment: dietary energy density and diet costs. *Am J Prevent Med* 27(Suppl):154–62.

Drewnowski, A. and Darmon, N. (2005a). The economics of obesity: dietary energy density and energy cost. *Am J Clin Nutr* 82(Suppl.):265S–73.

Drewnowski, A. and Darmon, N. (2005b). Food choices and diet costs: an economic analysis. *J Nutr* 135:900–04.

Drewnowski, A. and Levine, A.S. (2003). Sugar and fat – from genes to culture. *J Nutr* 133:829S–30.

Drewnowski, A. and Rolls, B.J. (2005). How to modify the food environment. *J Nutr* 135:898–99.

Fogg-Johnston, N. and Merolli, A. (2000). *Nutrigenomics: the next wave in nutrition research. Nutraceut World. (Online).* Available at: http://www.nutraceuticalsworld.com/articles/2000/03/nutrigenomics.php Accessed 26 February 2008.

French, S.A., Story, M. and Jeffery, R.W. (2001). Environmental influences on eating and physical activity. *Ann Rev Pub Hlth* 22:309–35.

Genelex (2006). *Nutritional genetics.* (Online). Available at: http://www.healthanddna.com/nutrigeneticstest.html Accessed 26 February 2008.

Genovations (2002a). *Physicians guide to clinical genomics.* (Online). Available at: http://www.genovations.com/home/clinician_overview.html Accessed 25 February 2008.

Genovations (2002b). *Genovations is the advent of truly personalized healthcare.* (Online). Available at: http://www.genovations.com/home/index.html Accessed 25 February 2008.

Ghosh, D., Skinner, M.A. and Laing, W.A. (2007). Pharmacogenomics and nutrigenomics: synergies and differences. *Eur J Clin Nutr* 61:567–74.

Gillis, C. (1999) 'Doomed' Newfoundlanders opt to eat, drink, and be merry. *National Post*, 12 Apr. p. A1.

Glanz, K., Basil, M., Maibach, E., Goldberg, J. and Snyder, D. (1998). Why Americans eat what they do: taste, nutrition, cost, convenience, and weight control concerns as influences on food consumption. *J Am Diet Assoc* 98:1118–26.

Government Accountability Office (GAO) (2006). *Nutrigenetic Testing: Tests Purchased from Four Web Sites Mislead Consumers.* (Online). Available at: http://www.gao.gov/new.items/d06977t.pdf Accessed 25 February 2008.

Health Canada (2001). *The population health template: key elements and actions that define a population health approach.* (Online). Population and Public Health Branch. Available at: http://www.phac-aspc.gc.ca/ph-sp/phdd/pdf/discussion_paper.pdf Accessed 26 February 2008.

Highfield, R. (2000). Selling science to the public. *Science* 289:59.

Hill, J.O. and Peters, J.C. (1998). Environmental contributions to the obesity epidemic. *Science* 280:1371–4.

Holistic Heal (2008). *Comprehensive Methylation Panel with Methylation Pathway Analysis #697.* (Online). Available at: http://www.holisticheal.com/store/product.php?productid=697&cat=124&page=1 Accessed 28 February 2008.

Hunter, D.J., Khoury, M.J. and Drazen, J.M. (2008). Letting the genome out of the bottle – will we get our wish? *N Engl J Med* 358:105–07

Johnson, R., Williams, S. and Spurill, I. (2006). Genomics, nutrition, obesity, and diabetes. *J Nurs Scholarship* 38:11–18.

Joost, H.-G., Gibney, M.J., Cashman, K.D., et al. (2007). Personalised nutrition: status and perspectives. *Br J Nutr* 98:26–31.

Kaput, J. (2005). Decoding the pyramid: a systems-biological approach to nutrigenomics. *Ann NY Acad Sci* 1055:64–79.

Kaput, J. and Rodriguez, R. (2004). Nutritional genomics: the next frontier in the postgenomic era. *Physiol Genomics* 16:166–77.

Kaput, J., Ordovas, J.M., Ferguson, L., et al. (2005). The case for strategic international alliances to harness nutritional genomics for public and personal health. *Br J Nutr* 94:623–32.

Kaput, J., Noble, J. and Hatipoglu, B. (2007). Application of nutrigenomic concepts to type 2 diabetes mellitus. *Nutr Metab Cardiovasc Dis* 17:89–103.

Kubik, M.Y., Lytle, L.A., Hannan, P.J., Perry, C.L. and Story, M. (2003). The association of the school food environment with dietary behaviours of young adolescents. *Am J Pub Hlth* 93:1168–73.

Labonte, R. and Penfold, S. (1981). Canadian perspectives in health promotion: a critique. *Hlth Educ* 19:4–9.

Laurance, J. (2005). Don't worry what diet to follow: willpower is the key ingredient. *The Independent*, 5 Jan. (Online) http://www.independent.co.uk/life-style/health-and-wellbeing/health-news/dont-worry-what-diet-to-follow-willpower-is-the-key-ingredient-489150.html Accessed March 19 2008.

Laverack, G. (2004). *Health promotion practice: power and empowerment*. Sage Publications, Thousand Oaks.

Lobstein, T. (1998). The common agricultural policy – a dietary disaster? *Consum Pol Rev* 8:82–7

MacGregor, H.E. (2005). Are the clues to diet success in your genes? *Los Angeles Times*, 11 Apr. p. F1.

Marteau, T.M. and Lerman, C. (2001). Genetic risk and behavioural change. *Br Med J* 322:1056–9.

Meijboom, F.L.B., Verweij, M.F. and Brom, F.W.A. (2003). You are what you eat: moral dimensions of diets tailored to one's genes. *J Agricult Environ Ethics* 16:557–68.

Mick, H. (2007). Personal genetic tests: genius or bogus? *Globe and Mail*, 26 Apr. p. L6.

Minnesota Department of Health (2006). Over the counter genetic tests and nutrition: what we know today. [Online]. Available at: http://www.health.state.mn.us/divs/hpcd/genomics/resources/fs/nutrigenomics.html Accessed February 25 2008.

Müller, M. and Kersten, S. (2003). Nutrigenomics: goals and strategies. *Nat Rev Genet* 4: 315–22.

Nestle, M. (2000). Changing the diet of a nation: population/regulatory strategies for a developed economy. *Asia Pacific J Clin Nutr* 9:S33–S40.
Newell, A., Zlot, A., Silvery, K. and Arail, K. (2007). Addressing the obesity epidemic: a genomics perspective. *Prevent Chronic Dis* 4:1–6.
Nisbet, M.C. and Mooney, C. (2007). Framing science. *Science* 316:56.
NuGO (European Nutrigenomics Organisation) (2004a). What is NuGO? (Online). Available at: http://www.nugo.org/everyone/24017 Accessed 26 February 2008.
NuGO (European Nutrigenomics Organisation) (2004b). Facts. (Online). Available at: http://www.nugo.org/facts Accessed 26 February 2008.
Nutrigenomics New Zealand (2004a). Tailoring New Zealand foods to match people's genes. (Online). Available at: http://www.nutrigenomics.org.nz/ accessed 26 February 2008.
Nutrigenomics New Zealand (2004b). About us. (Online). Available at: http://www.nutrigenomics.org.nz/index/page/26 Accessed 26 February 2008.
Ordovas, J.M. and Mooser, V. (2004). Nutrigenomics and nutrigenetics. *Curr Opin Lipidol* 15:101–08.
Ozdemir, V. and Godard, B. (2007). Evidence-based management of nutrigenomics expectations and ELSIs. *Pharmacogenomics* 8:1051–62.
Peregin, T. (2001). The new frontier of nutrition science: nutrigenomics. *J Am Diet Assoc* 101:1306.
Reidpath, D.D., Burns, C., Garrard, J., Mahoney, M. and Townsend, M. (2002). An ecological study of the relationship between social and environmental determinants of obesity. *Hlth Place* 8:141–45.
Reilly, P. and DeBusk, R.M. (2007). Ethical and legal issues in nutritional genomics. *J Am Diet Assoc* 108:36–40.
Ries, N. and Caulfield, T. (2006). First pharmacogenomics, next nutrigenomics: genohype or genohealthy?. *Jurimetrics J* 46:281–308.
Russo, G. (2006). Home health tests are 'genetic horoscopes'. *Nature* 442:497.
Salugen (2006) Individualized medicine. (Online). Available at: http://www.salugen.com/individualized-medicine.html Accessed 26 February 2008.
Sciona, Inc. (2007b). Mycellf: the science of you. (Online). Available at: http://mycellf.com/index.aspx Accessed 25 February 2008.
Sciona, Inc. (2007c). About sciona. (Online). Available at: http://www.sciona.com Accessed 26 February 2008.
Story, M., Neumark-Sztainer, D. and French, S.A. (2002). Individual and environmental influences on adolescent eating behaviours. *J Am Diet Assoc* 102(Suppl 3):S40–51.
Subbiah, M.T.R. (2007). Nutrigenetics and nutraceuticals: the next wave riding on personalized medicine. *Translat Res* 149:55–61.
Suracell (2007). How it works. (Online). Available at: http://www.suracell.com/how_it_works/core_nutrition.aspx Accessed 26 February 2008.
UC Davis NCHMD Center of Excellence for Nutritional Genomics (2007a). Home. (Online). Available at: http://nutrigenomics.ucdavis.edu/nutrigenomics/ Accessed 26 February 2008.
UC Davis NCHMD Center of Excellence for Nutritional Genomics (2007b). Information. (Online). Available at: http://nutrigenomics.ucdavis.edu/nutrigenomics/index.cfm?objectid=972D6E14-65B3-C1E7-053774E6C7AF510A Accessed 26 February 2008.
UK Human Genetics Commission (2007). More genes direct: a report on developments in the availability, marketing and regulation of genetic tests supplied directly to the public. (Online). Available at: http://www.hgc.gov.uk/client/document.asp?DocId=139&CAtegoryId=10 Accessed 26 February 2008.
Ursell, A. (2006). Family history holds clues to the best diet. *The Sunday Times Health Section* 3 Sept.
Wright, A.J., Wienman, J. and Marteau, T.M. (2003). The impact of learning of a genetic predisposition to nicotine dependence: an analogue study. *Tobacco Cont* 12:227–30.

Further reading

Abraham, C. (2005). Would you gaze into a genetic crystal ball? *Globe and Mail* 31 Dec. p. A1.

Bennett, P., Wilkinson, C., Turner, J., et al. (2007). The impact of breast cancer genetic risk assessment on intentions to perform surveillance behaviours. *J Genet Counsel* 1059:617–23.

Brody, J. (2008). No gimmicks: eat less and exercise more. *New York Times.*(Online). 1 January. Available at: http://www.nytimes.com/2008/01/01/health/nutrition/01brod.html Accessed 25 February 2008.

Canadian Biotechnology Strategy (2005). Introduction. (Online). Available at: http://www.biobasics.gc.ca/english/view.asp?x=715&mid=455 Accessed 26 February 2008.

Caulfield, T. (2000). Underwhelmed: hyperbole, regulatory policy, and the genetic revolution. *McGill Law J* 45:437–60.

CTV.ca (2007). Genes may determine caffeine risks: study. *CTV.ca* (Online). 7 March. Available at: http://www.ctv.ca/servlet/ArticleNews/story/CTVNews/20060307/coffee_genes_060307/20060307/ Accessed 26 February 2008.

Division of Cancer Prevention (2008) *Nutritional science research group*. (Online). Available at: http://prevention.cancer.gov/programs-resources/groups/ns Accessed 26 February 2008.

Elloit, R.M. and Johnson, I.T. (2007). Nutrigenomic approaches for obesity research. *Obesity Rev* 8(Suppl 1):77–81.

Genetic Health (2005). *The new nutrition gene*. (Online). Available at: http://www.genetic-health.co.uk/dna-test-services/nutritional-test.htm Accessed 26 February 2008.

GenSpec Labs (2006). *Genetically specific vitamins and supplements*. (Online). Available at: http://www.4genspec.com/ Accessed 26 February 2008.

Goddard, K.A.B., Moore, C., Ottman, D., Szegda, K.L., Bradley, L. and Khoury, M.J. (2007). Awareness and use of direct-to-consumer nutrigenomic tests, United States, 2006. *Genet Med* 9:510.

Institute of Medicine, Committee for the Study of the Future of Public Health (1998). The future of public health. National Academy Press, Washington, DC.

Interleukin Genetics (2006) Gensona genetic tests. (Online). Available at: http://www.ilgenetics.com/content/products-services/gensona.jsp Accessed 26 February 2008.

Jean Mayer USDA Human Nutrition Research Center on Aging (2008). Jean Mayer USDA Human Nutrition Research Center on Aging. (Online). Available at: http://hnrc.tufts.edu/1192109687036/HNRCA-Page-hnrca2ws_1192109688473.html Accessed 26 February 2008.

Morgan, S. and Hurley, J. (2002). Influences on the 'health care technology cost-driver'. (Online). Commission on the Future of Health Care in Canada. Available at: http://dsp-psd.pwgsc.gc.ca/Collection/CP32-79-14-2002E.pdf Accessed 26 February 2008.

NRIG (Nutrigenomics Research Interest Group) (2007). Mission. (Online). Available at: http://www.med.mun.ca/nrig/pages/02mission.htm Accessed 26 February 2008.

Public Health Agency of Canada (2008). Toward a healthy future: second report on the health of Canadians. (Online) Available at: http://www.phac-aspc.gc.ca/ph-sp/phdd/report/toward/index.html Accessed 26 February 2008.

Ries, N. (2008). Regulating nutrigenetic tests: an international comparative analysis. *Hlth Law Rev* 16:9–20.

Sciona, Inc. (2007a). Nutrigenomics. (Online). Available at: http://www.mycellf.com/nutrigenomics.aspx Accessed 25 February 2008.

Taylor, A. (2007) Urgent regulation needed for direct public gene tests, say experts. *BioNews*, (Internet). 10 December. Available at: http://www.bionews.org.uk/new.lasso?storyid=3662 Accessed 10 December 2007.

13

The Personal and the Public in Nutrigenomics

David Castle

OUTLINE

SUMMARY

Nutrigenomics research has been translated into direct-to-consumer applications. These commercial products and services focus on empowering people by individualizing and privatizing genomic technologies. The focus on personalization raises questions about whether public health

Nutrition and Genomics
ISBN: 978-0-12-374125-7

applications of nutrigenomics might be forthcoming. This chapter first reviews personalization of nutrigenomics as a knowledge translation focus of the Human Genome Project. Then, the potential for preventative nutrigenomic applications in a public health model is discussed, along with some potential nutrigenomic applications. The chapter concludes with some reflections on how public health nutrigenomics could be implemented and evaluated.

INTRODUCTION

Nutrigenomics presents an interesting, real-life case for considering persistent themes in human genomics about the relationship between individual people, the groups they belong to, their societies and the human population as a whole. As nutrigenomic science accumulates and matures, new information about the intersection of genetics and nutrition of individuals and groups is becoming available. Yet questions persist about whether future nutrigenomic scientific research and its translation into practical applications ought to focus on the needs of individuals, groups within populations or whole populations. These questions remain unresolved and the kind of evidence one would muster to decide them constantly changes as the science develops. At the same time, the field is being reshaped by a growing body of scientific literature, experience in the commercial domain, changing professional practice and regulatory scrutiny. The current trajectory is for nutrigenomics to continue down a path where translation of the science into applications focuses on individuals taking tests and making decisions more or less autonomously. Commercial activity in this vein has existed for the better part of a decade and the uptake of nutrigenomics through the practices of health care professionals, such as dietitians, will amplify the use of commercial services.

One might question whether continued personalization is the preferred path for nutrigenomics to follow. But are there other options for nutrigenomics where applications would benefit groups of people or perhaps the entire population of a country? There are really two questions mixed together here. The first is whether nutrigenomic applications should be preferentially developed for individual use, for use by groups of people or for broad use throughout populations. The second question is to an extent predicated on the first, which is whether the preferred applications ought to form endpoints that guide strategic decisions about research directions, funding and international collaborations.

Nutrigenomics is the study of the intersection of nutrition and genetics. With respect to the genetic aspect of nutrigenomics, the individual, not the public is the focus:

> Almost by definition, however, public health and genetics are incompatible. Public health is based on utilitarianism and paternalism. The benefit to society as a whole justifies coercive measures that outweigh individual rights. Consequently, a whole range of interventions – from immunization to isolation – may be justified. Genetics, on the other hand, has a completely different philosophical grounding. The intensely personal, inter-generational, and reproductive aspects of genetics have given rise to a professional ethos of non-directive counselling, autonomous decision-making, and individual rights – the very opposite of the approach of public health (Gerard et al., 2002).

With respect to the nutrition aspect, it is obviously the case that eating is at once a highly individual activity constrained and influenced by many social factors. These factors are often targets for change by those interested in the public health aspects of nutrition: public health nutrition aims to develop 'population-based strategies to promote good health through healthy diets' (Mensink and Plat, 2002). This guiding principle has been recognized for at least 50 years, since Keys et al. (1965) recognized that '... responsiveness to change in diet tends to be related to the intrinsic characteristics of the individual', and has been accepted community wisdom for very much longer. Broadly speaking, public health nutrition, among other objectives, seeks to reduce global and household food insecurity, improve the quality of diets and prevent and control nutritionally-related diseases (Darnton-Hill et al., 2004).

These considerations bring us to the question this chapter explores: is nutrigenomics strictly the science of, and for, individuals, or are there public health research questions and potential applications, as well? In this chapter, the potential for public health applications of nutrigenomics is considered as a different norm that can be adopted when thinking about knowledge translation of nutrigenomic science. The chapter begins by reflecting on the prioritization of the individual in nutrigenomics, before considering the alternatives. Next follows a discussion of how public health can be construed to give direction to the discussion of potential public health applications of nutrigenomics. In the final section, some practical and ethical challenges are considered to identify options for moving public health applications of nutrigenomics forward.

PERSONALIZATION – THE NORM OF NUTRIGENOMICS

Just before the release of the draft human sequences, a prominent group of commentators hoped the Human Genome Project (HGP) would help human genetics move beyond single disease genes to focus on gene–environment interactions with public health impact (Khoury et al., 2000). Publication of the draft sequences of the human genome (Lander, 2001; Venter, 2001) did raise hopes for rapid translation of the science into applications for the public, most of which took the form of personalized medicine (Bell, 2004), a field which has yet to fulfill the hype surrounding it (Khoury and Mensah, 2005). Of the three branches of environmental genomics – toxicogenomics, pharmacogenomics and nutrigenomics – it is the latter which has made the greatest strides in translating the 'basic' or 'discovery' science of the HGP into publicly-available, commercialized products and services. Setting aside issues about the overall quality of the applications of nutrigenomics science (Haga et al., 2003), it is inarguable that the translation of the HGP into these applications for the public has occurred. Initial adopters of nutrigenomics, as the Institute for the Future predicted in its early assessments of the market (Institute for the Future, 2001), find attractive the mix of personalization and new science. The end result is now a commercial market almost exclusively focused on personalized nutrigenomic testing. In the few years over which the science-to-business strategy evolved, personalization has come to guide nutrigenomic science in the search for nutrient–gene associations common enough to establish a cohort of affected individuals large enough to support commercially viable companies. This approach, however, has raised concerns in the scientific community that the kind of personalization being advocated focuses too narrowly on potentially paying consumers and not enough on the nutrigenomics of groups and populations. Social and ethical considerations about the ethos of individual autonomy dominating nutrigenomics suggest that greater benefits of nutrigenomics would be accrued if nutrigenomics were offered in a public health model.

The potential for a personalized diet is an intriguing prospect. Penders et al. (2007) observe, however, that the different meanings of personalization need to be considered. A literature review and series of interviews at nutritional genomics conferences showed that there were two different senses of 'personalized' circulating in discussions about nutrigenomics. One sense of personalization refers to 'individualization, in terms of diet, as the construction of either foods or dietary advice for the benefit of a *sole, unique* individual' (Penders et al., 2007). This is the definition used commercially, but, in research, a different definition of personalization is common. Considering that there are statistical limits to how fine-grained results of epidemiological analyses can be, personalization in nutrigenomics might not mean bespoke nutritional tailoring, but off-the-rack diet sizing. As one of their interviewees said,

> We do not tailor every article of clothing to the individual, we live comfortably with the fact that clothing sizes exist. This is the way in which I see genotyping. In the end we will be able to match a clothing size 42 to a genotype size 42. That means that we do not have to go down to the individual level, but we can also stay on the level of clothing size cohorts (Penders et al., 2007).

This conception of 'personalization' refers to 'personalization as categorization' in the sense that the utility in subdividing a population lies in finding groups that have relevant differences with respect to some objective. Therein lies a distinction with a difference. It may be the case that, '[d]espite the appeal of the "personalized diet", the practicalities of laboratory work make the "personal" less and less individualistic' (Penders et al., 2007). If that were true, and the 'sizing' of nutrigenomics is 'off-the-rack' rather than bespoke, then nutrigenomic testing would establish that *groups* of people ought to have *grouped* nutritional commonalities and differences. Equally, however, nutrigenomic testing could confirm that an individual tested has membership in a group, where individuals are clustered on the basis of at least one shared nutrigenomic trait.

The motivation for individual genetic testing for nutritional genomics is to determine whether a person has a genetic variant known to have a nutrient–gene interaction. Nutrient–gene association studies identify the frequency with which a polymorphism appears in a population and so any individual with a genetic variant has a trait relevant to them which, at the same time, identifies them as belonging to the group with the variation. Nutrigenomics has never been based on the premise that it would *uniquely* apply to individuals and never to groups. Personalization in the context of nutrigenomics is not radical. Besides, if it were, it would hardly make a viable commercial platform because genetics is not a science of individuals but of individuals within groups. Penders et al. are not saying that there is an irreconcilable tension between personalization meaning individuals and personalization meaning groups of individuals. They think these two 'notions of personalization' will continue to coexist, but with some ambiguity. From their standpoint, the most important issue is to determine what scientifically meaningful category one can use in nutrigenomics that will help to bridge these notions of personalization in a way that is useful for developing practical applications of nutrigenomics. Instead of a categorical conceptual difference, there is a degree of continuity along the continuum from individuals to populations, on which groups in the large, middle swath of the continuum is where all the debate is concentrated.

Penders et al. suggest that ethical issues arise in the act of categorizing individuals into groups defined by varying susceptibility to chronic disease. Individualism can raise other ethical issues in nutrigenomics suggests Chadwick (2004). Chadwick argues that a major ethical issue confronting nutrigenomics lies in the difference between 'prevailing ethical paradigms, which are predominantly concerned with individualism and choice, rather than with the common or public good'. In addition to the traditional bioethical issues about the relationship between individuals participating as research subjects, their autonomy relative to the greater good and public benefit of individual contributions to research databases, Chadwick thinks there is an important, but largely unrecognized, ethical issue in nutrigenomics regarding the difference between personalized uses of nutrigenomics and public applications. She suggests genetic screening

> . . .is typically defined as the determination of the prevalence of a gene in an asymptomatic population or population group, where for any given individual there is no reason to believe that he or she has the gene in question. It is normally contrasted with the genetic testing of an individual for whom there may be some reason to think he or she is at risk (Chadwick, 2004).

Chadwick thinks that having symptoms motivates genetic testing in medical contexts and rightly points out that in current applications of nutrigenomics, people might be tested presymptomatically. There are some exceptions – testing for genetic variants associated with hemochromatosis might be triggered by symptoms – but nutrigenomics applications tend to focus on simple tests for reducing chronic disease risk in the susceptible. There is no barrier to using nutrigenomic tests as part of a diagnostic in a symptomatic person, but the potential for inroads into mitigating disease in people susceptible to diet-related chronic disease is a novel value proposition to scientists and the private sector. Chadwick argues that an ethical framework focused on individual autonomy is problematic in its own right and problematic for its consequent lack of support for the public good. In other words, ethical theory is biased toward individual autonomy and so defenses of nutrigenomic applications that directly or implicitly laud individualism steer nutrigenomics away from serving the interests of groups or the public in general. Chadwick does not, however, make the case for the superiority of the public interests over the individual interest and she does not discuss the meaning and extent of the division between autonomous individual ethics and public or common good ethics as these concepts apply to nutrigenomics.

PUBLIC HEALTH AND NUTRIGENOMICS

Despite the shortcomings in Penders et al. and Chandwick's arguments, the point that motivates their positions is worth considering: it is an important question whether nutrigenomics has been dominated by an ethos of individual autonomy or personalization, with less attention to public health applications of nutrigenomic science. On the face of it, being able to translate nutrigenomic research results into accessible products and services for the greatest number of people is preferable. Does that mean public health approaches automatically trump individual-based applications? Tempting as it might be to seize upon this conclusion, there are important considerations about what would constitute a viable and useful public health application of nutrigenomics. Nutrition is obviously a major determinant of health and public health measures can focus on changing dietary patterns, with or without a genetic component. To use the genetic component effectively in public health nutrigenomics, '[i]n terms of preventing chronic disease in more susceptible individuals and differing responses to dietary modification, it is likely that there will be some increased targeting of individuals and sub-populations, to make more efficient dietary recommendations' (Darnton-Hill et al., 2004). The authors might have meant 'more effective', not just 'efficient', dietary recommendations. If nutrigenomics allows for greater precision in targeting dietary interventions to groups who will benefit the most, less time and effort will be expended on improving the health of groups and the effects ought to have greater impact.

Public health applications of nutrigenomics rest on the fundamentals of public health. The Institute of Medicine's 20-year-old definition of public health remains foundational to the entire field:

> Public health is what we, as a society, do collectively to assure the conditions in which people can be healthy (Committee for the Study of the Future of Public Health, 1988).

This definition is usefully expanded by Childress et al. who define public health as being:

> [P]rimarily concerned with the health of the entire population, rather than the health of individuals. Its features include an emphasis on the promotion of health and the prevention of disease and disability; the collection and use of epidemiological data, population surveillance, and other forms of empirical quantitative assessment; a recognition of the multidimensional nature of the determinants of health; and a focus on the complex interactions of many factors – biological, behavioural, social and environmental – in developing effective interventions (Childress et al., 2002).

From these definitions, a contrast can be made between clinical medicine and public health. In medicine, the focus is generally on symptomatic individuals who seek a diagnosis, symptom alleviation, treatment and cure. This is not the focus of public health, which seeks to 'understand the conditions and causes of ill-health (and good health) in the populace as a whole. It seeks to assure a favourable environment in which people can maintain their health' (Gostin, 2005).

Generally, public health definitions strive to differentiate between clinical medicine interventions to benefit the individual and population level interventions to benefit the public. Considering the roles and responsibilities of the state, the public and individuals, the intent is to differentiate between what is normally expected or required of autonomous individuals and what the state might enact for the benefit of the public where limitations on some individual's autonomy has to be balanced against benefits to the population. To grasp better what public health means in terms of autonomy and justice, the public is often characterized in three ways: as the 'numerical population' which is the targeted group; the 'political public' which refers to the government and public agency; and the public in the broad, collective sense or 'communal public' (Childress et al., 2002). Efforts to improve the health of the public through preventative measures or health promotion can target a numerical public, have public agency and generate outcomes where the overall health of the community improves. What is meant by the outcome of public health can be a property of the community taken as a whole or an aggregated property of individuals. In the first case, herd effects of public health programs for vaccinations have community-wide effects because they lower the incidence of communicable disease for the public as whole. Aggregated individual health benefits, for example through genetic disease screening, improve the on-average health of the public by benefitting a targeted group.

Gostin points out that while much of the impetus behind what he describes as a 'renaissance' in interest in public health comes from the threat of infectious diseases and bioterrorism, the other important 'stimulus' comes from recognizing the burden of chronic diseases on health systems (Gostin, 2005). As he points out, the significant behavioral elements in chronic diseases lead societies to consider ethical and legal options for altering people's behavior for healthier outcomes:

> Central to these debates are arguments about personal responsibility, which leads to disturbing social fault lines based on politics and ideology. Are individuals responsible for their own behavior or are they influenced by social and cultural factors that can, and should, be changed?

How one characterizes the domain of public health initiatives is quite important. Unlike Gostin, Rothstein, for one, thinks 'public health' does not include state interventions that would lead to health promotion, in other words, treating the presymptomatic (Rothstein, 2002). Gostin, on the other hand, believes it does, because there is room for a broader conception of public health ethics that ought to be endorsed and anchored in public health interventions (Gostin, 2004). This includes chronic diseases that involve important dietary and lifestyle components with which public health officials must grapple. Inclusion of chronic disease within the public health mandate certainly broadens the range of complicated issues facing public health officials. As recent disputes about the obesity epidemic suggest, for example, there are radically different causal accounts regarding the most important causes of obesity. These range from psychological observations about behavior and willpower, to the impact of food environments, built environments, advertising, sociocultural associations about food and state food policy. Consensus views about the cause of epidemic obesity are seemingly not on the horizon and so it is impossible, if not reckless, to try to pin blame on any one individual or corporate actor.

Nevertheless, public health is not just an observational science based on epidemiology and biostatistics, it is also a manifestation of public policy and political will. Despite the fact that accounts of the causes of public health problems and their solutions might be incomplete and underdetermined by data, states are responsible to some degree for the health and well-being of citizens, irrespective of where the state's policies lie on a continuum from interventionist to libertarian ideologies. Consequently, attempts to enact different mechanisms for achieving public health goals will receive varying support and will be variously justified in different jurisdictions. For example, attempting to alter built, information or socio-economic environments to address obesity will be successful to varying degrees, as will more direct approaches such as taxation, statutory and non-statutory regulation and deregulation (Ries, 2008). Furthermore, when one considers a narrower field in which public health interventions might be possible, such as nutrigenomics, states can draw on a smaller set of strategies.

Public health measures generally attempt to improve the quality of life and longevity of a population by reducing incidence of disease and injury and by making treatments more effective and accessible. Implementation of measures generally seeks to attain three, desirable attributes: community involvement and civic responsibility; a strong prevention orientation; and social justice (Gostin, 2005). While individuals are the ultimate beneficiaries of public health measures, the overall benefit is a property of the population as a whole and is not simply decomposable to the benefits of individuals. In the case of vaccines, for example, childhood vaccination programs certainly benefit individual children, but their value to a society as a whole is a population-attributable benefit.

In addition to seeing a positive health outcome from a public health intervention, there are other, ethical considerations. Kass (2001) suggests a number of elements that ought to be considered when evaluating the ethical dimension of public health initiatives. The public health goal itself must be defensible before any initiative is launched and the program must be capable of achieving the stated goals. Any new burdens for individuals as a result of adopting the plan must not be unduly onerous and certainly must not wholly detract from the accrued benefit. Burdens ought to be minimized, alternative approaches must be sought and the burdens should preferably not be assigned to any one group if the population as a whole receives the benefits (Kass, 2001). These principles are not wholly dissimilar from Upshur's view that public health interventions are justified just so long as they do not violate the harm principle, involve unduly restrictive or coercive means, where there is reciprocity and where the intervention is transparent to the public (Upshur, 2002).

These criteria provide a framework for thinking about the general contours of nutrigenomics in a public health context that help to identify potential candidates for public health nutrigenomics. Above all else, the ethical criteria set out by Upshur and Kass set the bar high for thinking about benefits to the public that are fairly distributed and not based on a biased distribution of costs of the public health measure. Those interested in implementing a public health measure in nutrigenomics must meet a burden of proof based on a cost-benefit-risk analysis. Second, the planned initiative needs to show that the benefit will be received by a large enough number of people so that the aggregated benefit for the population warrants public intervention. Third, a public health measure ought to be realized within the normal channels of intervention discussed above by Gostin. With these criteria in mind, the matter therefore becomes one of finding candidate nutrigenomic applications that fit.

PUBLIC HEALTH APPLICATIONS OF NUTRIGENOMICS

Public health can include interventions focused on health promotion in presymptomatic individuals, it can focus on targeted groups or more broadly and it can be implemented through a variety of techniques including information campaigns or targeted screening. Interventions also ought to be fairly distributed, cost effective, justifiable for a public health intervention and developed using normal intervention pathways. The difficulty in applying these criteria to nutrigenomics is that early genetics research focused first on single disease genes and then more common polymorphisms. The field of nutrigenomics is only now starting to examine the contribution that multiple, common polymorphisms make to disease susceptibility.

Take, for example, the case of alcohol consumption, a common subject of public health messages about the potentially harmful effects of overconsumption. As a bioactive component of many people's diets, alcohol is also believed to have other, genotoxic effects. Ishikawa and colleagues have studied alcohol metabolism and cancer, focusing on the variants of two enzymes, alcohol dehydrogenase-2 (ADH1B) and acetaldehyde dehydrogenase (ALDH2) (Ishikawa et al., 2007). While alcohol is not considered a carcinogen on its own, its metabolic by-product, acetaldehyde, is. Genetic variants of ADH1B and ALDH2 affect the body's ability to clear alcohol and acetaldehyde and prolonged levels of acetaldehyde are associated with higher than normal levels of micronuclei – a common measure of genome instability. This research follows upon other work by Ishikawa that showed that cytochrome P450 variants, in conjunction with ALDH2 and ADH variants, raised individual susceptibilities to the carcinogenic effects of alcohol metabolites (Ishikawa et al., 2006). These results could be useful in a screening program, especially if in combination with other related single-nucleotide polymorphisms (SNPs), for example those associated with folate metabolism and colorectal cancer.

The concept of genome stability through nutrition is gaining ground. Essentially, the idea is that genetic damage occurs throughout one's lifetime and that nutrition plays a role in providing micronutrients necessary for DNA repair – thus contributing to genome stability. Fenech compares the impact of low folate on lymphocytes in laboratory settings with an exposure of more than ten times the annual safe limit of X-rays (Fenech, 2005a). Potentially carcinogenic in low intake, calcium, folate, retinol and several other micronutrients must be studied further to learn if common polymorphisms in metabolic pathways frequently lead to greater disease susceptibility (Fenech, 2005b). Recommended daily intake of common micronutrients might have to change in light of nutrigenomic research (Fenech, 2002), 'Genomic Health Clinics' might have to assist specific subgroups with micronutrient intake and the potential is there for 'massive numbers of health-conscious consumers to be able to assess directly the effect of their dietary and nutritional supplement choices on their genome and that of their children' (Fenech, 2005a). If genome stability proves to be a fruitful strategy for nutrigenomics to follow, it could lead to population health measures such as information campaigns for revised recommended daily intake of various nutrients that play a direct role in DNA repair.

Nutrients appear to have a role in providing genome stability from conception. There is growing evidence that nutritional environments are major sources of epigenetic events (Ross, 2003; Davis and Uthus, 2004; Mathers, 2005). One candidate for public health measures is among the more interesting applications of nutrigenomics: metabolic imprinting during development. Gluckman et al. report that nutritional status of gestating rodents affects metabolic imprints of their offspring (Gluckman et al., 2007). The relationship between maternal nutritional status and offspring's propensity to be obese, exhibit hyperinsulinemia and also hyperphagia is directionally dependent, suggesting that the effect is irreversibly epigenetic. If this were true of humans, it could have profound implications for how women ought to be advised about their diets during pregnancy. Chavez and Munoz de Chavez consider maternal epigenetics as being a worthy case of public health nutrigenomics (Chavez and Munoz de Chavez, 2003). Their stance is that compared to the evolution of the human genome, human's diets have changed rapidly in the last few ten thousands of years.

Other recent work focuses on the rate of change of human genomes, arguing that there is strong evidence that human evolution accelerated partly in response to environmental factors like diet (Hawks et al., 2007), but Chavez and Munoz de Chavez think dietary changes outstrip human evolution, leading to environmental pressure under which some genotypes are better adapted than others. For example, moving to an agrarian lifestyle and cultivating a broader food base has revolutionized human life in an evolutionary short time frame and even more drastic has been the impact of rapidly changing dietary environments such as that experienced by Native Americans converting to European-based dietary traditions during the course of a few generations. In either case, these are very short evolutionary time frames. Some individuals might not respond as well as others to rapidly changing nutritional environments, making adaptive epigenetic changes all the more important. If Gluckman is right that we can discern which changes, through maternal imprinting, would make us better adapted to our nutritional environments, Chavez and Munoz de Chavez hypothesize that we might be able to eat in adaptive ways. The idea of nutrigenomic fetal epigenetic programming has been discussed elsewhere (Burdge et al., 2007) and it raises the possibility that developmental origins of significant chronic diseases could be addressed through maternal nutritional patterns (McMillen and Robinson, 2005). In a public health model, this information could be given easily to women consulting physicians and midwives.

IMPLEMENTATION AND EVALUATION

The Third European Nutrigenomics Conference in 2006 included discussion of the potential for nutrigenomics to make inroads into public health nutrition. Summarizing the conference discussion, Lampe observed the following:

> It is easy to imagine how better characterizing the intricacies of nutrient–gene interactions will further our ability to more accurately pinpoint nutrient requirements, as well as define the contribution of other dietary constituents to disease prevention and health. On the other hand, promotion of good health through nutrition and the primary prevention of nutrition-related illness in the population (i.e. public health nutrition) require the application of existing nutrition, behavior and public health knowledge and the consequent adoption of established dietary recommendations by a population. Fully characterizing dietary exposures and biological responses to them using the available omics technologies and systems biology techniques will improve nutrition knowledge, but the support needs to be there for the next steps – applying and adopting this knowledge (Lampe, 2006).

Lampe is correct to point out that integration of genomics into existing public health measures will require advancements in public health as well as in public health applications of genomics and genetics. Without these advancements, it is difficult to think that programs of public health nutrigenomics would be feasible.

As Khoury and Mensah point out (Khoury and Mensah, 2005), the Institute of Medicine (IOM) identified genomics as one of eight cross-cutting priorities in their 2003 report (Institute of Medicine, 2003). Khoury and Mensah argue in support of the IOM perspective and point out that there is a clear path for integration of genomics into public health. Family history data are a routine aspect of medical consultations and it ought to be possible to integrate these data into genomic databases. In support of this kind of initiative, they argue that population-based research and databases should be developed for genomics and public health. This approach would give population prevalence data sorely needed to evaluate the potential impact of modifying environmental factors to reduce risk. It would be predictive because the genomic and pedigree data could be used in risk algorithms. Second, they propose that genomic applications in public health should have an evidence-base for health promotion and disease prevention. Moving away from a patient-centered to a public health approach would likely involve fewer high risk genetic factors and interventions, favoring instead susceptibility markers and advice generalized to groups. Finally, no public health application of genomics will be complete without institutional capacity and human competence building, although as Guttmacher et al. (Guttmacher et al., 2007) and Farrell (see Chapter 8) point out, genetics and genomics training is still not prevalent in North American curricula for health care providers. Genomics and genetics are far from being routine in health care (Foster et al., 2006).

Integrating family history data collection with genomics databases provides a potentially cost-effective and feasible approach to public health uses of genomics and this information could support public health nutrigenomics initiatives. Genetic screening is another potential technique to consider, although it presents some unique challenges in the public health context. In addition to demonstrating that the test is accurate according to established criteria (such as Good Laboratory Practices and Clinical Laboratory Improvement Amendments), the test should lead to a clinically validated outcome. Given persistent concerns about the public taking advantage of tests but lacking the support necessary to interpret and act upon the information, it has been suggested that classification of tests according to the impact of the information and the potential to take action should be created (Burke et al., 2001). Presumably, counseling proportionate to the gravity of the health risks (see Chapter 5) and the potential for action would be a necessary component of any comprehensive public health system utilization of genetic tests. Furthermore, privacy and non-discrimination safeguards ought to be in place and implemented in a manner so that unduly heightened individual concerns about privacy do not trump other individuals' rights of access to testing and to their genetic information (Hodge, 2004).

Several years ago, Khoury et al. (2000) discussed the role of effective communication about genetics in creating effective public health uses of genetics and genomics. They argued that, for effective translation of genomics, three important messages need to be communicated to the public. The first is that disease is really a function of gene–environment interactions, second, that population research, of the sort described above, is crucial to demonstrate clinical validity and utility of genetic testing, and third, that genetic information supports targeted use of interventions. These messages present a communication problem identified by the authors:

> Such messages could lead to a better appreciation of how behaviors can reduce disease risk. Risk reduction in the face of genetic susceptibility is complex. It will require additional research in risk communication and the behavioral sciences to understand how to safely and effectively bring about the changes needed to improve health among those who have genetic susceptibility (Khoury et al., 2000).

Here the authors identify a significant challenge, which is to communicate effectively about susceptibility to chronic disease and how one would go about managing that risk. If there are no known interventions that would reduce risk, their conclusion is that '[u]ntil such interventions are available for any given genetic trait, identification of genetic susceptibility may do more harm than good' (Khoury et al., 2000). This perhaps overstates the case somewhat, since the issue is not the underlying genetic trait that demands an intervention, but the condition that the gene–environment gives rise to. Public health nutrigenomic advice would not be 'change your genes' but 'change your nutritional environment to change your gene expression'.

One remaining issue to consider is the role of evidence in deciding which public health nutrigenomic applications are supported by available evidence. Intermediate sources of evidence, such as biomarkers of exposure and metabolism, can provide insight about how a public health program is working, but outcomes of reduced chronic disease mortality and morbidity attributed to public health nutrigenomics might take longer than conventional measures normally allow.

Wilson discusses the central problem facing public health officials, which is to balance the need for decision-making against information uncertainty (Wilson, 2005). The issue is that, in cases of uncertainty about the prospects of implementing a program, or foregoing it, officials neither want to waste scarce resources, nor do they wish to make decisions that worsen the health of the public.

> All these policy responses demonstrate the need for formulating policy in advance of clear evidence about risk. They put forth the question as to what exactly is the appropriate role of evidence in the policy process? When should policy-makers consider the evidence of causation substantial enough to act on it? Under what circumstances would it be premature to address a potential risk? (Wilson, 2005).

Wilson argues that there are essentially two paradigms: one is an evidence-based approach that tries to establish causation using norms of science drawn from clinical medicine, and the other approach involves using the precautionary principle. The evidence-based approach looks first to see if there is a statistically significant relation between two events (e.g. the diet pill fenfluramine and pulmonary hypertension). The next step is to establish the nature of the association, while ruling out possible sources of error. Finally, a credible biological hypothesis explaining the validated statistical association needs to be generated. This model, however, has limitations. As Wilson points out, it does not tell health officials if they are able, or not, to act on preliminary data correlating two events. It is also difficult to achieve a silver standard by developing a case-controlled study, let alone the gold-standard randomized control trial. Wilson cautions that misuse of evidence, as in the case of the 1998 *Lancet* study in which autism and MMR (measles, mumps, rubella) vaccinations were linked, can have harmful ripple effects.

In light of these likely knowledge limitations, decision makers must decide if a public health initiative has high or low benefit and high or low risk to the public if implemented. If the right kind of evidence is lacking, public health officials might have to adopt a precautionary approach. Wilson refers to the Wingspread definition of the precautionary principle:

> Where an activity raises threats of harm to the environment or human health, precautionary measures should be taken even if some cause and effect relationships are not fully established scientifically. [. . .] In this context the proponent of the activity, rather than the public bears the burden of proof (Ashford et al., 1998).

Not all countries endorse the Wingspread definition directly in their policies. Canada, for example, takes a precautionary *approach* that is derived from this definition. It guides decision makers in cases where there is a gap in evidence and requires of health officials that they take into consideration five principles each of precautionary application and measure before taking a decision (Health Canada, 2003). Wilson suggests that public health officials can find some middle ground between a clinical evidence-based approach and precaution by adopting a modeling strategy that establishes how widespread and harmful the exposure (or susceptibility in the case of nutrigenomics) is, the harm caused to individuals of not proceeding with a public health measure and availability of strategies to minimize economic and health costs. In the context of nutrigenomics this suggests that where the risks of implementing a public health application of nutrigenomics is low, but the benefits could be relatively high because many people are susceptible, officials might find a middle ground accommodation for the technology's application.

CONCLUSION

Public health applications of genomics, in the form of nutrigenomics and pharmacogenomics, lie in the future. Farmer and Godard (2007) say of public health applications of pharmacogenomics that, '[t]he translation of genomic knowledge into clinical practice reveals the possibility of changing not only the way research in public health is conducted, but also how subsequent interventions may be carried out'. Genomics research is rapidly changing, with the advent of genome-wide association studies (The Wellcome Trust Case Control Consortium, 2007) and deep phenotyping (Tracy, 2008) and developments in how individuals and groups are understood in terms of metabolic types (Kaput, 2008). Nutrigenomic interventions promise to reduce the risk of chronic diseases by targeting those who might be most susceptible but, currently, these tests are offered privately to individuals. If the benefits of prevention are significant, it may be financially rewarding to offer preventative tests through public health channels (Morgan et al., 2003).

References

Ashford, N., et al. (1998). Wingspread Statement on the Precautionary Principle Available from www.gdrc.org/u-gov/precaution-3.html.

Bell, J. (2004). Predicting disease using genomics. *Nature* 429:453–6.

Burdge, G.C., Hanson, M.A., Slater-Jefferies, J.L. and Lillycrop, K.A. (2007). Epigenetic regulation of transcription: a mechanism for inducing variations in phenotype (fetal programming) by differences in nutrition during early life?. *Br J Nutr* 97:1036–46.

Burke, W., Pinsky, L.E. and Press, N.A. (2001). Categorizing enetic tests to identify their ethical, legal and social implications. *Am J Med Genet* 106:233–40.

Chadwick, R. (2004). Nutrigenomics, individualism and public health. *Proc Nutr Soc* 63: 161–6.

Chavez, A. and Munoz de Chavez, M. (2003). Nutrigenomics in public health nutrition: short-term perspectives. *Eur J Clin Nutr* 57(Suppl 1):S97–100.

Childress, J.F., Faden, R.R., Gaare, R.D., et al. (2002). Public health ethics: mapping the terrain. *J Law Med Ethics* 30:170–8.

Committee for the Study of the Future of Public Health, Division of Health Care Services, Institute of Medicine (1988). The Future of Public Health. National Academy Press, Washington, DC.

Darnton-Hill, I., Margetts, B. and Deckelbaum, R. (2004). Public health nutrition and genetics: implications for nutrition policy and promotion. *Proc Nutr Soc* 63:173–85.

Davis, C.D. and Uthus, E.O. (2004). DNA methylation, cancer susceptibility, and nutrient interactions. *Exp Biol Med (Maywood)* 229:988–95.

Farmer, Y. and Godard, B. (2007). Public health genomics (PHG): from scientific considerations to ethical integration. *Genomics, Soc Po* 3:14–27.

Fenech, M. (2002). Micronutrients and genomic stability: a new paradigm for recommended dietary allowances (RDAs). *Food Chem Toxicol* 40:1113–17.

Fenech, M. (2005a). The genome health clinic and genome health nutrigenomics concepts: diagnosis and nutritional treatment of genome and epigenome damage on an individual basis. *Mutagenesis* 20:255–69.

Fenech, M. (2005b). Low intake of calcium, folate, nicotinic acid, vitamin E, retinol, b-carotene and high intake of pantothenic acid, biotin and riboflavin are significantly associated with increased genome instability – results from a dietary intake and micronucleus index survey in South Australia. *Carcinogenesis* 26:991–9.

Foster, M.W., Royal, C.D.M. and Sharp, R.R. (2006). The routinisation of genomics and genetics: implications for ethical practice. *J Med Ethics* 32:635–8.

Gerard, S., Hayes, M. and Rothstein, M.A. (2002). On the edge of tomorrow: fitting genomics into public health policy. *J Law Med Ethics* 30(Suppl):173–6.

Gluckman, P.D., Lillycrop, K.A., Vickers, M.H., et al. (2007). Metabolic plasticity during mammalian development is directionally dependent on early nutritional status. *Proc Natl Acad Sci USA* 104:12796–800.

Gostin, L.O. (2004). Law and ethics in population health. *Aus NZ J Publ Hlth* 28:7–12.

Gostin, L.O. (2005). The core values of public health law and ethics. In *Public Health Law and Policy in Canada* (T.M. Bailey, T. Caulfield and N.M. Ries, eds.). LexisNexis, Markham.

Guttmacher, A.E., Porteous, M.E. and McInerney, J.D. (2007). Educating health-care professionals about genetics and genomics. *Nat Rev Genet* 8:151–7.

Haga, S.B., Khoury, M.J. and Burke, W. (2003). Genomic profiling to promote a healthy lifestyle: not ready for prime time. *Nat Genet* 34:347–50.

Hawks, J., Wang, E.T., Cochran, G.M., Harpending, H.C. and Moyzis, R.K. (2007). Recent acceleration of human adaptive evolution. *Proc Natl Acad Sci* 104:20753–8.

Health Canada (2003). The application of precaution – a framework for the application of precaution in science-based decision-making about risk. Ottawa: Government of Canada. Available at http://www.pco-bcp.gc.ca/docs/information/publications/precaution/precaution_e.pdf.

Hodge, J.G. Jr. (2004). Ethical issues concerning genetic testing and screening in public health. *Am J Med Genet* 125C:66–70.

Institute for the Future (2001). The future of nutrition: consumers engage with science. Cupertino. Available at http://www.iftf.org/node/925.

Institute of Medicine (2003). Who will keep the public healthy? Educating public health professionals for the 21st century. National Academies Press, Bethesda.

Ishikawa, H., Miyatsu, Y., Kurihara, K. and Yokoyama, K. (2006). Gene-environment interactions between alcohol-drinking behavior and ALDH2 and CYP2E1 polymorphisms and their impact on micronuclei frequency in human lymphocytes. *Mutat Res* 594:1–9.

Ishikawa, H., Ishikawa, T., Yamamoto, H., Fukao, A. and Yokoyama, K. (2007). Genotoxic effects of alcohol in human peripheral lymphocytes modulate by *ADH1B* and *ALDH2* gene polymorphisms. *Mutat Res* 615:134–42.

Kaput, J. (2008). Nutrigenomics research for personalized nutrition and medicine. *Curr Opin Biotechnol* 19:1–11.

Kass, N.E. (2001). An ethics framework for public health. *Am J Publ Hlth* 91:1776–82.

Keys, A., Anderson, J.T. and Grande, F. (1965). Serum cholesterol response to changes in the diet. III. Differences among individuals. *Metabolism* 14:766–75.

Khoury, M.J. and Mensah, G.A. (2005). Genomics and the prevention and control of common chronic diseases: emerging priorities for public health action. *Prev Chronic Dis* 2:A05.

Khoury, M.J., Thrasher, J.F., Burke, W., Gettig, E.A., Fridinger, F. and Jackson, R. (2000). Challenges in communicating genetics: a public health approach. *Genet Med* 2: 198–202.

Lampe, J.W. (2006). For debate: investment in nutrigenomics will advance the role of nutrition in public health. *Cancer Epidemiol Biomarkers Prev* 15:2329–30.

Mathers, J.C. (2005). Nutrition and epigenetics – how the genome learns from experience. *Br Nutr Foundation Nutr Bull* 30:6–12.

Mensink, R.P. and Plat, J. (2002). Post-genomic opportunities for understanding nutrition: the nutritionist's perspective. *Proceedings of the Nutrition Society* 61:401–4.

McMillen, I.C. and Robinson, J.S. (2005). Developmental origins of the metabolic syndrome: prediction, plasticity, and programming. *Physiol Rev* 85:571–633.
Morgan, S., Hurley, J., Miller, F. and Giacomini, M. (2003). Predictive genetic tests and health system costs. *Can Med Assoc J* 168:989–91.
Penders, B., Horstman, K., Saris, W.H.M. and Vos, R. (2007). From individuals to groups: a review of the meaning of 'personalized' in nutrigenomics. *Trends Food Sci Technol* 18:333–8.
Ries, N.M. (2008). Piling on the laws, shedding the pounds? The use of legal tools to address obesity. *Hlth Law J* (Supplement):101–26.
Ross, S.A. (2003). Diet and DNA methylation interactions in cancer prevention. *Ann NY Acad Sci* 983:197–207.
Rothstein, M. (2002). Rethinking the meaning of public health. *J Law Med Ethics* 30:144–9.
The Wellcome Trust Case Control Consortium (2007). Genome-wide association study of 14,000 cases of seven common diseases and 3,000 shared controls. *Nature* 447:661–78.
Tracy, R.P. (2008). 'Deep phenotyping': characterizing populations in the era of genomics and systems biology. *Curr Opin Lipidol* 19:151–7.
Upshur, R.E.G. (2002). Principles for the justification of public health intervention. *Can J Publ Hlth* 93:101–3.
Wilson, K. (2005). Risk, causation and precaution: understanding policy-making regarding public health risks. In *Public Health Law and Policy in Canada* (T.M. Bailey, T. Caufield and N.M.R. Ries, eds.). LexisNexis, Markham.

Further reading

Lander, E.S., et al. (2001). Initial sequencing and analysis of the human genome. *Nature* 409:860–921.
Venter, J.C., et al. (2001). The sequence of the human genome. *Science* 291:1304–51.

Food Styles and the Future of Nutrigenomics

Michiel Korthals

OUTLINE

Nutrition and Genomics
ISBN: 978-0-12-374125-7

SUMMARY

New scientific and technological developments do not happen in a social vacuum but in contexts structured by people's interests, claims and values. In the case of nutrigenomics, many normative assumptions are at play. Nutrigenomics researchers have specific ideas about food and health, some of which are present in the social context of food consumption and production in which nutrigenomics and its applications will develop. Food performs several 'functions' in our lives and our associations with food can be summarized into different 'food styles'. Understanding how these interact in the case nutrigenomics shows that current conceptions of nutrigenomics do not exhaust the potential domain for nutrigenomics applications. The chapter concludes with suggestions about how to broaden nutrigenomic applications into a greater number of food styles and the consequent ethical and cultural effects.

INTRODUCTION

New scientific and technological developments, like nutrigenomics, do not happen in a social vacuum but in a pre-structured territory where parties (stakeholders) have interests, claims, values and moral positions. Stakeholders' attitudes and positions, one way or another, will influence scientific and technological developments. At the same time, by engaging with new technological developments, stakeholders will adapt their preferences, interests and values, both voluntarily and non-voluntarily. Several core values are implicated in nutritional genomics: health, the meaning of food, and eating and responsibility. Each of these values is subject to constant change.

In this chapter, normative assumptions concerning concepts of health and food evinced by many nutrigenomics researchers are first outlined. The current view is strongly oriented toward personalized health, prevention of calculable risks and personal responsibility. Second, the social sector of food consumption and production in which nutrigenomics and its applications will develop are analyzed. Several functions of food are distinguished, including nutritional, aesthetic, social and cultural functions, and the shifting relationship of food with medicine is commented on. Different styles and cultures of eating (or 'food styles') are then assessed, in particular with respect to their ethical status and their relationship with the development of nutrigenomics. Some of these food styles align more with current normative assumptions of nutrigenomics and are therefore likely receptive to the information and options nutrigenomics offers. Nutrigenomics could be helpful in the context of other food styles as well, but only with changes in assumptions involved in the current research agenda. Both the public and private sectors – including government, civil society organizations and food companies – have responsibility in broadening the normative assumptions of nutrigenomics. The general conclusion is that knowledge of the healthfulness of food consumption, derived through nutrigenomics research and its applications, can heighten the pleasures of eating and be culturally and ethically acceptable in any of the food styles.

CURRENT NORMATIVE ASSUMPTIONS ON HEALTH AND FOOD IN NUTRITIONAL GENOMICS

Presently, among the different factors that influence health, diet is seen as a primary factor in increasing the risks of certain diseases. This implies a new conception of health and food. A core assumption of personalized nutrition is that food is no longer seen as positive (i.e. having pleasurable features) or as neutral, but as having side effects that disadvantage human health in the long run (like cardiovascular disease, obesity and cancer). Nutrigenomics holds that people should choose their diet on the basis of their genetic profile and that diet-related diseases should be detected as early as possible to facilitate prevention through dietary modification. According to Kaput and Rodriquez (2004):

> the tenets of nutritional genomics are: 1) common dietary chemicals act on the human genome, either directly or indirectly, to alter gene expression or structure; 2) under certain circumstances and in some individuals, diet can be a serious risk factor for a number of diseases; 3) some diet-regulated genes (and their normal, common variants) are likely to play a role in the onset, incidence, progression, and/or severity of chronic diseases; 4) the degree to which diet influences the balance between healthy and disease states may depend on an individual's genetic makeup; and 5) dietary intervention based on knowledge of nutritional requirement, nutritional status, and genotype (i.e. 'individualized nutrition') can be used to prevent, mitigate, or cure chronic disease. (See also Müller and Kersten, 2003)

In a recent study, Komduur et al. (2008) came to the conclusion that the *first* normative assumption found in current nutrigenomics representations is that values regarding food are exclusively explained in terms of health. Health is construed narrowly as disease prevention, but implies a broad and unremitting responsibility for active prevention. Pursuit of health (i.e. disease prevention) through dietary choices is seen as a constant matter of concern. The *second* normative assumption maintains that health can and should be explained in relation to quantifiable and calculable risks. This notion holds the promise of control and the minimization or even banishment of risks. Health is not seen as fate but as something to produce by healthy dieting. The most important function of food, then, is in addressing 'predisease' risks. The *third* (normative) assumption is that disease prevention by minimization of risks with diets is ultimately in the hands of the individual (although perhaps on the basis of group membership) and that this kind of risk information should motivate personal disease prevention. These normative assumptions on health and food, focused on personal nutrition, dominate the current view of nutrigenomics. Together they define a conception of health that can be called 'preventive and individualized health'. However, nutrigenomics incorporates other potentialities and health conceptions, which could be of use in, for example, public or collective health and food programs.

FUNCTIONS OF FOOD: NUTRITIONAL, CULTURAL, SOCIAL AND AESTHETIC ASPECTS

Research and commercialization in nutrigenomics occurs within contemporary and varied food cultures. How can nutrigenomics align with these food styles and their disparate views on functions of food? Food studies distinguish between the physiological, aesthetic and communicative aspects of eating and consider social, cultural and moral functions. The food we eat says something about our taste (aesthetic meaning), it has a function in establishing and maintaining social integration in a group (social meaning) and it serves to identify status and other differences between groups (cultural meaning). Lastly, the food you eat tells something about the kind of person you are, about the type of life you wish to lead and, of course, about the fair or unfair distribution of wealth (moral meaning). The physiological function relates strictly to survival and the body's needs to maintain itself.

In some eating cultures, such as the dominant eating culture in the USA, food is considered mainly for its health qualities and not primarily as a product to be enjoyed (Levenstein, 1993). The influence of the food sciences is one reason for this, but a deep-rooted puritanism is likely another principal factor. In other cultures, such as France and Italy, taste is dominant and food consumption is linked with pleasure. Social food studies have supplied a flood of data and theories to show that taste is a socially determined factor that is dependent on the place and position of the consumer in the social network (Bourdieu, 1984). The social function means that individuals come together and confirm their mutual relationships during meals. Food is part of the cultural lifestyle in which people grow up, from which they may later distance themselves (or not) and derive their identity.

Both the quantity and quality of food products serve as a means of communication among individuals and groups. The quantity, for example, can indicate how hospitable someone is and to what extent others can count on him or her in times of food scarcity and other needs. Large food quantities imply a hospitable, open and uncomplicated meal structure, such as in the traditional American kitchen where, in principle, anyone can sit down at the table at any time of the day. With small quantities, cooking can be very exquisite, fewer people may be invited and the sequence of courses is more regulated (Douglas, 1982). The French kitchen is a good example of this. Eating a great deal or little can have the same function. Consuming much food can be a sign that you appreciate the hospitality and are prepared to do a favor in return. An altogether different example is the Japanese rice culture, which confirms formal and hierarchical relationships and is an expression of such relationships (Visser, 1986; Ohnuki-Tierney, 1993). A dinner with all sorts of rice products, including sake or rice wine, can be a formal theatrical performance in which positions are sounded out or confirmed.

The differentiating function involves groups distinguishing themselves from each other through consumption of specific foods and avoidance of others. Mintz (1996) claims: 'The way bread tastes, the way dough is prepared and baked, unites people culturally'. But the taste of bread and the method of baking also differentiate people from each other. Differences in eating practices may be attenuated in a less stratified, more egalitarian society, but societies with significant differences in rank and social class reveal enormous variation in cuisines. New food products often lead to great cultural controversies, especially when novel products do not integrate easily into existing nutritional views. For example, religious groups in the 16th and 17th centuries objected to potatoes and tobacco, 'newfangled' food products mentioned nowhere in the Bible. In the 19th century, anti-slavery activists protested against sugar under the slogan, 'Eating sugar means eating negroes' (Mintz, 1996).

The paradox of the desire for differentiation is that early adopters of a food trend distinguish themselves from others, but wider adoption of a successful trend eventually dissolves the delineation. In that way, the elitist bohemian culture of the romantic poets of the 19th century, with their great emphasis on outdoor hiking, traveling and culinary adventures, became a more or less general trend in the 20th century (Campbell, 1987).

A complex set of considerations influences individuals' food choices and these considerations are typically different from those that determine choice and use of pharmaceuticals. At least four factors determine general food preferences: genetic factors, social and cultural factors, bodily history and personal historical factors (Korsmeyer, 2002). *Genetic factors* are generally not changeable and influence taste (e.g. bitter and sweet) and other reactions to food (e.g. allergies). *Social and cultural factors* shape values and practices regarding food and eating at all levels, including individual, family, community, region and nation. *Bodily history* influences tongue and mouth, but also nose and ear, all of which engage with food and eating. Finally, *personal historical factors* involve family history and an individual's adoption of particular eating choices and practices. Moreover, in actually having a meal, additional factors play a role, like the present conditions of the meal, the present condition of the body and the actual pleasure and displeasure in tasting. Gustatory taste and food choice are strongly intentional and cognitive: it matters what one eats and to know what one eats. They comprise various sensory perceptions: visual, smell (olfactory), tongue (gustatory) and auditory perceptions. The fact that taste is cognitive means taste can be acquired and unlearned. Through learning from those around us, a yellowish substance with a rotting smell may become known as a delicious cheese and associated with a good taste experience.

LIFESTYLES AND FOOD STYLES

Food has many more meanings than simply its significance in terms of health. Food is part of an individual's identity and shapes group membership or sets groups apart in terms of ethnicity, race, nationality, class, individuality and gender (Cockerham et al., 1997). In short, food choices are part of a lifestyle. Sociologists such as Giddens (1991), Beck (1990) and Schulze (1993) have shown that we live in a late or postmodern society in which individualization has led to a multiplicity of lifestyles that also represent multiple – sometimes unconventional – forms of mutual solidarity and socialization. The characteristics of lifestyles, after all, are shared socially. Lifestyles are constantly in a state of flux, to a significant degree because of technological and scientific developments, such as computer technology, modern means of communication, or new means of transport. In his magnificent empirical and conceptual study of late-modern lifestyles, Schulze (1993) differentiates three general categories:

- *upper culture*, whose typical features include a predilection for classical music, 'real' literature and French cuisine
- *mass culture*, which favors popular music, television game shows and fast food
- *action culture*, which identifies with rock music, thrillers and ethnic foods.

These three categories result in at least five lifestyles, each implying a specific food style:

- the *integration* lifestyle: stands for sociability (against upper culture and for mass culture) and regular consumption of traditional foods
- the *harmony* lifestyle: stands for formal and proper behavior and a sense of security (for upper culture); frequent consumption of fast food

- the *upper level* lifestyle: stands for the pursuit of better and higher things in life (for upper culture); frequent consumption of slow food
- the *self-realization* lifestyle: more artistic, visits bars, is into eco-tourism (for upper culture and action culture); focused on slow food and urban community farming
- the *amusement* lifestyle: focused on TV stars, is into fitness and gambling (mass culture and action culture); frequent consumption of fast food and functional foods.

What an upper level lifestyle adherent would call good and interesting (such as an academic lecture), the amusement lifestyle adherent would call boring or arcane. What the upper level lifestyler would call primitive or trivial, the amusement lifestyler would call exciting and judge the upper level lifestyler as arrogant or aloof. Wolfe (1994) distinguishes roughly the same categories, asserting that each category is represented by approximately one fifth of the total US population. Most European countries have five or six different lifestyles with generally an equal number of adherents (Brunso, 1995; Rozin et al., 1999).

Lifestyles do not fit within a hierarchy. Since there is more than one cultural elite, the lifestyle of the upper class is no longer exemplary for the lower class and a dynamic mechanism between a high and a low culture is missing. Such dynamism might only work *within* one of the lifestyles mentioned, between trendsetters and followers within that style. Schulze (1993) also shows that much 'shopping' takes place, with frequent switching between lifestyles (see also Shields, 1992). Adoption of specific lifestyles and food styles is rooted in many factors, but empirical relationships between motivating factors and individual choices are not always relevant from an ethical point of view. However, no one maintains the same lifestyle all the time and people in late-modern societies regularly switch between lifestyle segments. Even fervent adherents of the upper level lifestyle do not always eat in fashionable settings but will occasionally consume a McDonald's hamburger. This means that the concept of lifestyle is not seen as a description but as a heuristic tool to categorize people's choices in life.

These different lifestyles comprise different conceptions of food and health. In general, we can distinguish different ways of relating to food, such as the dichotomy expressed in whether one eats to live or lives to eat. Moreover, food styles have different attitudes towards food advice, to traditional and new ways of cooking and eating (neophobia or neophilia), to eating outside the home and to gendered forms of eating (Warde, 1997; Miele and Murdoch, 2002). Roughly speaking, these styles concentrate on traditional (regional or national) food, health food, natural (organic) food, international (cosmopolitan) food or fast food (Korthals, 2004). Many consumers do not adopt only one style, but are rather eclectic. Two individuals may choose the same kind of foods, but their motivation may be totally different. For example, an organic food style may be motivated by personal health considerations or concern for animal welfare. The five most common food styles are:

- In the integration lifestyle, *traditional* (*national*) *food* plays a crucial role and the relationship with disease prevention is less significant
- In the harmony lifestyle, *fast food* and fast (e.g. microwave) cooking is a distinctive food choice and preventive health is only relevant if it can be solved in a fast way, e.g. by supplements
- The upper lifestyle is oriented toward *cosmopolitan and high quality food*: much effort is invested in good (seen as artisanal) ingredients and in getting good food advice from international gourmets. 'Slow food', the food movement started in Italy but now expanded all over the world in opposition to fast food, often attracts its adherents from this lifestyle. Health (individualized and preventive) plays a minor role
- The self-realization lifestyle is connected with *natural and organic food* and food movements, very often concerned with health in an alternative, heterodox and holistic way
- The amusement lifestyle has strong connections with *fast food* and is willing to invest time, energy and money into preventive health and (functional) food.

HOW WILL CURRENT FOOD STYLES REACT TO NUTRIGENOMICS AND THE NEW RELATIONSHIP BETWEEN FOOD AND DRUGS?

Viewed historically, the boundaries between food and drugs are often unclear, indeed, the ancient Greek thinker, Hippocrates, counseled: 'let food be your medicine!' (Korthals, 2004). In pre-Enlightenment Europe, the ideas of the Greek physician, Galen, had a dominant influence on cooking, stressing the intimate relationship between pleasure and health. Galen's ideas were simple and specific: every living being is a particular mixture of dry and moist and hot and cold and health means a good balance between the four (Temkin, 1973).

With the rise of medical and food sciences, however, new boundaries were drawn between food and drugs, as summarized in Table 14.1.

Nutrigenomics provokes new considerations of these distinctions between food and medicine. As said, some food styles associate food more strongly with preventive and individualized health (Cain and Schmid, 2003) and would likely be more receptive to services and products based on nutrigenomics conceptualized as personalized nutrition, for example

TABLE 14.1 Boundaries between foods and drugs

	Food	Drugs
Consumption	For all people	Only for persons with disease
Negative side effects	Should not be present because long-term consumption is not optional	Acceptable because drugs are usually taken on a temporary basis
Motivations for consumption	Motivated by many factors	Motivated by desire to be cured or protected
Availability	Widely available	Available only with medical prescription
Regulatory review	For safety only	For safety and efficacy

the fast food food style. Other food styles, especially those that adopt a traditional, social, culinary or slow food approach, are less likely to adopt nutrigenomics. Carlo Petrini (2002), the founder of the Slow Food movement and a fierce defender of the social and cultural aspects of enjoying food, approvingly quotes Madame Sevigny: 'Health is enjoying the other enjoyments. When the other enjoyments are taken away, we live longer, but we lose our health'. The Slow Food movement, which is gaining considerable momentum in the western world, clearly does not subscribe to the narrow definition of health propagated by personalized nutrition and nutrigenomics.

In the same vein, the traditional and cosmopolitan food styles have objections against the identification of food and medicine. Proponents of these styles would say that society is not a hospital, meaning that health should not be the all-determining value in food choices. Food contributes to the values of society (traditional food style) or to the conversation of humankind (cosmopolitan style) and if food is only produced with a view to health, or more specifically, with a view to disease prevention, these other values may be lost (Korthals, 2004). The English Food Ethics Council (FEC) views the exclusive orientation on health in food choices as a transformation of society into a hospital (FEC, 2005) and scholars warn about the medicalization of daily life (see e.g. Crawford, 1980).

PEACEFUL COEXISTENCE AND FAIR REPRESENTATION OF FOOD STYLES

A key ethical issue in this landscape of food styles concerns how these styles can flourish without harming one other and their adherents. The concept of 'peaceful coexistence of cultures of eating' is helpful here. This concept fits into a liberal ethical framework, in which individuals have the right to choose freely a life and food style, as long as it does not harm the public interest (Korthals, 2008).

What is the implication of this peaceful coexistence of life and food styles for the public (governmental, civil society) and private sector (market, food companies)? Governments should strive for fair representation of food styles in policy, research and the regulation of markets and not privilege one particular food style over others. Governments should enable the food styles to flourish and interact freely with each other and, in particular, governments should enable citizens and consumers to move freely from one food style to another (e.g. taxes and subsidies should not make food connected with a particular style disproportionately expensive or cheap compared to the common foods of other food styles). Governments should also protect citizens and consumers against harm by fixing standards for food safety, animal welfare and sustainability in food production. Companies should be responsive to market preferences of consumers for different food styles and take into account the fair representation of food styles. Food science and food research should pay as much attention to food as health as for food as taste and other aesthetic values. Nutrigenomics, in short, should broaden its conception of healthy food, which requires a reorientation of the current research agenda towards all the food styles.

NUTRIGENOMICS AND FAIR REPRESENTATION OF FOOD STYLES

Nutrigenomics, both in its current form as personal nutrition and as a potential for public health and collective nutrition (e.g. battling chronic diseases), may conflict with some food styles by emphasizing that food is the paramount means to achieve health. First, food is part of daily existence and pressure to eat for health, which nutrigenomics advocates, may begin to permeate shopping, cooking and eating. The connection between food and health in itself may not be harmful, except when the emphasis on health may increase the tendency in society to overestimate the significance of health in a narrow sense. This overestimation of health could mean that health becomes a social norm to which everyone should comply (Beck-Gernsheim, 1994). This norm may become so time-consuming that individuals will have no time for other activities. Crawford first raised this concern in the 1980s, using the term *healthism* to describe the characterization of health as a 'super value'. With health as a super value, the pursuit of the good life becomes a search for health (Crawford, 2006). Health becomes a norm and the failure to achieve health or to seek it is then seen as a failure to embrace life. It can be predicted that many, maybe even the majority of the population, will not live up to this value.

Second, nutrigenomics in its current form emphasizes the prevention of 'prediseases'. It encourages people to eat for the future and could bring them to disregard their present needs, like socializing on the basis of food or culinary needs. The idea of prevention has more fundamental assumptions. In addition to privileging the future over the present, it also assumes that all prediseases will develop to full diseases if dietary and other lifestyle changes are not adopted. This assumption, however, is false as people may live with severe risks without any symptomatic health problems or because prediseases simply do not always evolve into full-fledged diseases.

Third, nutrigenomics may increase personal responsibility and decrease social responsibility for health, which implies that public authorities have fewer obligations to care for the health of the population. Some food scientists and food industrialists contend that genetic testing and personalized nutrition are means of individual empowerment and should replace public health nutrition policies (German and Watzke, 2004; Koelen and Lindström, 2005). Similarly, health insurers and government departments advocate that consumers should take more responsibility for health, such as the controversial UK Department of Health report, *Choosing Health* (2004). In counterpoint, the UK Public Health Association (UKPHA, 2005) paper, *Choosing Health or Losing Health*?, argues public health should be the main target for governmental policies. Where individual responsibility is the main organizing principle, population health overall suffers as health disparities between socioeconomic classes are exacerbated (Darnton-Hill et al., 2004). Lang and Heasman (2005) also make a strong case for the primacy of public over individualized health nutrition politics: 'Targeting whole populations provide governments with better chances of public health success, whereas targeting "at risk" individuals could be socially divisive. This does not mean, as is sometimes assumed, everyone eating the same or a bland diet, but moving overall dietary behaviour *en masse* in a healthier direction'. Many argue for harnessing genomics towards public health priorities (Khoury and Mensah, 2005). In other words, nutrigenomics puts anew the issue of 'who is responsible for health' on the public agenda. The chapter will conclude with a few reflections on that issue.

MORAL RESPONSIBILITY FOR HEALTH IN DIFFERENT FOOD STYLES

What is moral responsibility?

Moral responsibility applies in respect of areas over which a person exerts control, but also, in a certain sense, to things that happen to an individual and are seemingly beyond one's control. Terrible things may happen when a person drives a car after drinking too much and, although the individual is not in control anymore, he or she can and should know beforehand that drinking and driving do not go together. Individual knowledge about potential consequences of behavior means that one can be called morally negligent even where one has not directly and intentionally caused harm.

Moral responsibility for health

Most philosophers would agree that the responsibility to stay healthy lies in the first instance with the individual. The individual is the one who controls his or her body (an argument from possession) and knows best what the body needs to stay healthy (an argument from epistemology). Even in cases of drunkenness or types of addiction, one could say at least the second argument still holds. Exclusive emphasis on disease prevention – for example, through technological developments like nutrigenomics – actually represents a form of disempowerment, because it often invalidates personal knowledge people have of their body and promotes reliance on the advice of others. In the words of Kant (1784):

> Laziness and cowardice are the reasons why such a large part of humanity, even long after nature has liberated it from foreign control (*naturaliter maiorennes*), is still happy to remain infantile during its entire life, making it so easy for others to act as its keeper. It is so easy to be infantile. If I have a book that is wisdom for me, a therapist or preacher who serves as my conscience, a doctor who prescribes my diet, then I do not need to worry about these myself. I do not need to think, as long as I am willing to pay.

Emphasizing individual responsibility for health undermines public health policies and health care institutions, which require social responsibility and public resources. What is then the relationship between the role of individual and social responsibility in the case of nutrigenomics? As long as health remains one of the considerations in food choice and supply and the multiple functions of food stay in place, there is good reason to reconsider the intricate relationship between the two (Korthals, 2004). When personal nutrition on a modest scale is implemented, it indeed produces a shift towards more personal responsibility, but does not eliminate collective responsibility altogether. Indeed, in the post-modern time, individuals perceive more and more their body as a resource to design for a happy, healthy and long life. As Giddens states: '... the body is becoming a phenomenon of choice and action' (Giddens, 1991).

There is one complicating factor here and that is that individuals can only assume more responsibility for health if they can cope with the fundamental uncertainties and ethical concerns of a science-based food and health policy. Giddens (1991) observes, '[t]he capability of adapting freely chosen lifestyles, a fundamental benefit generated by a post-traditional order, stands in tension, not only with barriers to emancipation, but with a variety of moral dilemmas'. In connection with the three normative assumptions of nutrigenomics, one can distinguish at least three intriguing dilemmas. One of them concerns the *health dilemma*: should the value of health, redefined as a super value, overshadow all other values of life? The second is the *prevention dilemma*: should prevention of foreseeable diseases go so far that present needs are neglected? The third one is the *individualization dilemma*: should individualization of responsibility go so far that the force of collective responsibility diminishes considerably?

RESPONSIBILITIES REVISITED

Nutrigenomics and its implications for food, health and the body create dilemmas and ambiguities for consumers and regulators. With nutrigenomics we do not live in the risk society but in the dilemma society as the complex relationship between food, genetics and health becomes even more complex. Generic health advice becomes unsatisfactory in light of growing knowledge of individual variation in response to foods and susceptibility to disease. Further research in nutrigenomics is needed to support tailored recommendations and deliver the promise of nutrigenomics for individuals, in policy measures and in ethically acceptable research agendas.

An *ethics for consumers* can receive inspiration from Kant's adage 'don't let your doctor *dictate* your diet' which, in a certain sense is right, but does not assist in navigating new connections between food and drugs. The advice of influential food writer Michael Pollan (2008) to eat only food one's grandmother would recognize is impossible to follow in a new situation where boundaries between food and drugs are increasingly blurred. Physicians and other health care experts have a traditional role in advising patients but traditional ethical concepts like 'informed choice' and 'individual responsibility' are not adequate for food ethics (Chadwick, 2004). 'Informed choice' has a function in protecting individuals against professionals, like physicians. When trust between individuals and experts is at stake, however, and when the collective implications of the implementation of personalized nutrition are so huge, deliberative ways of decision-making should be tried in which all stakeholders have to readjust their opinions and interests according to the ongoing discussion. This implies that the ethics of protection of the individual against the apparently mighty powers of state and professionals is not fruitful; more adequate is an ethics of participation in continuous discussion on the intricacies of nutrigenomics.

There is a second implication for consumers. Consumers need an ethics of dealing with dilemmas and uncertainties in the sense of identifying and selecting between important, major ones and minor, unimportant ones. There are no general tools or general guidelines to deal with that process of selection, but there are general procedures like consultations, deliberations and exchange of stories and life narratives. The main thing here is to find out commonalities and particularities in your own life and that of other affected ones. Some ethical support can be given by Putnam's distinction between common sense doubt, meaning the selection of more and less certain cases, and philosophical doubt, i.e. the radical denial of all certainties (Putnam, 1995):

> We have to remind ourselves of the distinction between common sense doubt and philosophical doubt. Finishing in believing in something is not really a human possibility. Criticism cannot be a reason for universal scepticism. The fact that sometimes we are wrong is not a reason to really doubt every particular conviction (Putnam, 1995).

Common sense doubt shifts between different types of doubt that are more or less realistic. Informed choice is not a good ethical concept in this context because it does not say anything about selecting the more certain and less certain recommendations and the incorporation of well-established health considerations into one's diet and personal health. We need ethical categories that assist consumers in making these selections; i.e. we need categories of consultation and deliberation and of learning processes in which consumers can bring their life stories (narrative input) and make them more robust using offered health services. This ethics of cooperation between consumers and professionals and of sharing life experiences implies that consumer groups should try to bridge the gap between production and consumption of food and new technologies and to incorporate the new health considerations into their food style together with the main stakeholders like governments and food industry (Liakopouklos and Schroeder, 2003).

With respect to *the ethics for governments and their food policy*, the uncertainties in the gray zone between food and health are best tackled not by prohibiting the gray zone but by regulating these on a social and technological level (Korthals, 2004) and taking into account the fair representation of food styles. Governments should organize the research agenda of nutrigenomics in a democratic way, not exclusively oriented towards personalized nutrition, but oriented towards the prevention of common illnesses, common conditions and chronic diseases. There is a role to play for public health nutrigenomics, directed at general risk profiles. Governments should regulate the three sectors, food, drugs and the emerging gray zone of functional foods or nutraceuticals that are at the boundary of foods and drugs.

CONCLUSION

The highly individualized approach to nutrigenomics reinforces narrow conceptions of food for health. But is this approach conducive to the good life as envisioned by diverse food and lifestyles? The narrow perspective of nutrigenomics ought to be broadened to be applicable in the wider context in which people cook and eat according to different food styles. Nutrigenomics can be applied across all food styles. For example, health considerations can be combined with taste in the slow food style, as is done in the branch of nutrition science concerned with molecular gastronomy (This, 2006).

Along the lines of Kant, it can be argued that personalized nutrition and personal health responsibility can mean disempowerment in its emphasis on food in a medicalized, disease prevention context. There are, however, some reasons in favor of personalized nutrition, as long as it does not push public health nutrition aside. Also, from a utilitarian standpoint, there is place for personalized nutrition in situations where unhealthy lifestyles cost too much and individuals bear some responsibility for those costs.

If nutrigenomics remains constrained to the three normative assumptions discussed in this chapter, the science will produce a shift towards emphasizing food for health and drive more personal responsibility for health than most life and food styles now accept. These two outcomes seem rather untenable. If, however, nutrigenomics broadens its scope to represent all the food styles, it can produce knowledge, services and products useful to people who follow diverse lifestyles. Finally, the pleasures of eating in all food styles can be heightened by the knowledge of its healthfulness as revealed through an expanded scope of nutrigenomics science.

References

Beck, U. (1990). *Risk society*. Sage, London.

Beck-Gernsheim, E. (1994). Gesundheit und Verantwortung im Zeitalter. In *Riskante Freiheiten* (U. Beck and E. Beck-Gernsheim, eds.). Suhrkamp, Frankfurt, pp. 316–35.

Bourdieu, P. (1984). *Distinction*. Harvard University Press, Cambridge.

Brunso, K. (1995). Development and testing of a cross-culturally valid instrument: food-related life style. *Adv Consum Res* xxii:475–80.

Cain, M. and Schmid, G. (2003). From nutrigenomics science to personalized nutrition: the market in 2010. The Institute for the Future, CAL.

Campbell, C. (1987). *The romantic ethic and the spirit of modern consumerism*. Routledge, London.

Chadwick, R. (2004). Nutrigenomics, individualism and public health. *Proc Nutr Soc* 63:161–66.

Cockerham, W., Rutten, A. and Abel, Th. (1997). Conceptualizing contemporary health lifestyles. *Sociol Q* 38:321–42.

Crawford, R. (2006). Health as a meaningful social practice. *Health* 10(4):401–20.

Crawford, R. (1980). Healthism and the medicalization of every day. *Int J Hlth Serv* 10:3.

Darnton-Hill, I., Margetts, B. and Deckelbaum, R. (2004). Public health nutrition and genetics: implications for nutrition policy and promotion. *Proc Nutr Soc* 63:173–85.

Douglas, M. (1982). Food as a system of communication. In *The active voice*. Routledge, London, pp. 81–118.

FEC (Food Ethics Council) (2005). Getting personal: shifting responsibilities for health. Food Ethics Council, Brighton.
German, J.B. and Watzke, H.J. (2004). Personalizing foods for health and delight. *Comprehens Rev Food Sci Food Safety* 3:145–51.
Giddens, A. (1991). *Modernity and self-identity*. Stanford, London.
Kant, I. (1784). What is enlightenment? In *Kant's Critique of practical reason and other writings in moral philosophy*. Translated and edited by L.W. Beck, pp. 263–92. Chicago University Press, Chicago.
Kaput, J. and Rodriquez, R.L. (2004). Nutritional genomics: the next frontier in the postgenomic era. *Physiol Genomics* 16:166–77.
Khoury, M.J. and Mensah, G.A. (2005). Genomics and the prevention and control of common chronic diseases: emerging priorities for public health action. *Prev Chronic Dis* (serial online). Available from http://www.cdc.gov/pcd/issues/2005/apr/05_0011.htm.
Koelen, M.A. and Lindström, B. (2005). Making healthy choices the easy choices: the role of empowerment. *Eur J Clin Nutr* 59(Suppl 1):S10–S16.
Komduur, R.H., Korthals, M. and Te Molder, H. (2008). The good life: living for health and a life without risks? On a prominent script of nutrigenomics. *British Journal of Nutrition*, doi:10.1017/S0007114508076253
Korsmeyer, C. (2002). *Making sense of taste*. Cornell University Press, Ithaca, NY.
Korthals, M. (2004). *Before dinner: philosophy and ethics of food*. Springer, Dordrecht.
Korthals, M. (2008). Ethics and Politics of Food; Toward a Deliberative Perspective. *Journal of Social Philosophy* 39(3):445–63.
Lang, T. and Heasman, M. (2005). *Foodwars*. Earthscan, London.
Levenstein, H. (1993). *Paradox of plenty: a social history of eating in modern America*. Oxford University Press, New York.
Liakopoulos, M. and Schroeder, D. (2003). Trust and functional foods. New products, old issues, poiesis & praxis. *Int J Technol Assess Ethics Sci* 2:41–52.
Miele, M. and Murdoch, J. (2002). Fast food/slow food: standardising and differentiating cultures of food. In *Globalization, localization and sustainable livelihoods* (R. Almas, ed.). Ashgate, Aldershot.
Mintz, S. (1996). *Tasting food, tasting freedom*. Baecon Press, Boston.
Müller, M. and Kersten, S. (2003). Nutrigenomics: goals and strategies. *Nat Rev Genet* 4:315–22.
Ohnuki-Tierney, E. (1993). *Rice as self*. Princeton University Press, Princeton.
Petrini, C. (2002). *The pleasures of slow food: celebrating authentic traditions, flavors and recipes*. Chronicle Books, San Francisco, CAL.
Pollan, M. (2008). *In defence of food*. Penguin, London.
Putnam, H. (1995). *Pragmatism: an open question*. Blackwell, Oxford.
Rozin, P., Fischler, C., Imada, S., Sarubin, A. and Wrzesniewski, A. (1999). Attitudes to food and the role of food in life in the USA, Japan, Flemish Belgium and France: possible implications for the diet–health debate. *Appetite* 33:163–80.
Temkin, O. (1973). *Galenism: Rise and Decline of a Medical Philosophy*. Cornell UP, Ithaca.
Schulze, G. (1993). *Die Erlebnisgesellschaft. Kultursoziologie der Gegenwart*. Campus, Frankfurt.
Shields, R. (ed.) (1992). *Lifestyle shopping: the subject of consumption*. Routledge, London.
This, H. (2006). *Molecular gastronomy*. Columbia University Press, New York.
UKPHA (2005). *Public Health Association, choosing health or losing health?* Public Health Association, London.
Visser, M. (1986). *Much depends on dinner*. MacMillan, New York.
Wolfe, D.B. (1994). Targeting the Mature Mind American Demographics, 16, 32-35.
Warde, A. (1997). *Consumption, food and taste, culinary antinomies and commodity culture*. Sage, London.

Further reading

Castle, D., Cline, C., Daar, A.S., Singer, P.A. and Tsamis, C. (2006). *Science, society and the supermarket: the opportunities and challenges of nutrigenomics*. Wiley, Cambridge.

Grunert, K. (1997). Food related lifestyle: results from three continents. *Asian Pacific Adv Consum Res* 2:64065.

O'Neill, O. (2002). Public health or clinical ethics: thinking beyond borders. *Ethics Int Affairs* :16.

UK Department of Health (2004). *Choosing Health*. Department of Health, London.

CHAPTER

15

Epilogue: Future Directions

David Castle and Nola M. Ries

OUTLINE

SUMMARY

This *Epilogue* crystallizes the book's diverse content in a set of recommendations that provide future direction for the field of nutrigenomics. These recommendations are derived from an analysis about the most important issues in the current cycle of nutrigenomics and focus on the topics central to this book: science, ethics, law, regulation and communication. Chapter authors were asked to identify dominant and emerging issues in their topic area and to propose how the issues could be addressed through research, business activity, regulation, communication, policy setting or political action.

Nutrition and Genomics
ISBN: 978-0-12-374125-7

The editors and chapter authors met in person in January 2008 at Banff, Canada, for a workshop to discuss content of draft chapters, focusing particularly on identification of the key issues and potential solutions. Following the workshop, authors further refined their chapters with the benefit of comments and suggestions from our cross-disciplinary meeting. The summary presented here is a distillation of the workshop discussion of these issues and solutions, together with paraphrasing and reorganization of the authors' written submissions.

DEVELOPING AND TRANSLATING NUTRIGENOMIC SCIENCE

A number of recommendations address tools to aid advancement of nutrigenomic science and translation into beneficial applications for the public. To begin, limitations in current study design are a dominant concern. As Ordovas and Tai indicate in this book's first chapter, these include the size of the populations studied and the methods used to evaluate dietary intake. Researchers should address these problems through better experimental methodologies and protocols. In addition, Krul and Gillies advocate for better characterization of the nutrients studied in nutrigenomics. Because they are generally not well characterized, the literature contains inconsistent and discordant data. The scientific community should be encouraged to improve data reporting standards and funding agencies and journal editors should impose higher standards as a condition of funding and publication. They further propose criteria for a 'minimal information for a dietary protein' standard, but the idea for benchmarking bioactive components of food can apply to molecules other than proteins.

Krul and Gillies also identify a need to develop conventions for research protocols and biobanks. Genotyping costs are decreasing and the technology is increasingly available. With routine genotyping in clinical nutrition studies, researchers could better explore the impact of genetics on the interaction of lifestyle factors with nutrient metabolism. Having genetic information available can help researchers further define and differentiate the study population, particularly if the study population is small to the point of running the risk of sample bias undermining the study's conclusions. Like Ordovas and Tai, Krul and Gillies recognize that these issues relate directly to study design and suggest that if more studies had metabolomic data clearly defining 'responder phenotypes' in intervention trials, subsequent genotyping is always possible.

Getting the science right through improved study design and through standardization, consolidation and sharing of data are important first steps in developing the scientific base from which to translate science into useful applications. Often, talk of the applications of nutrigenomics immediately provokes the direct-to-consumer controversy discussed elsewhere in this book. Yet there is an important intermediary, the food companies, which are generally mute on the issue of nutrigenomics, but obviously have a keen interest in good quality science upon which to base significant investment decisions. They need to know who will benefit from potential new products that promise nutritional pre-emption of disease and why. The other intermediary, of course, are the small firms selling nutrigenomic products and services directly to consumers. As Gill points out, the entire field of nutrigenomics must show that the science is credible to build consumer trust. The success of nutrigenomics depends on relevant and useful applications for the public. For Krul and Gillies, this amounts to having 'actionable knowledge' that leads to demonstrable health benefits and, as many authors in this book point out, that knowledge must be communicated to the public in a way that is truthful and comprehensible.

REGULATING NUTRIGENOMIC TESTS AND HEALTH CLAIMS

Were it not for small firms attempting to translate nutrigenomic science into practical applications, the regulatory environment for nutrigenomics would not be the controversial issue that it is. Perhaps this is an overstatement, in the sense that there are other kinds of direct-to-consumer (DTC) genetic tests on the market other than nutrigenomic tests. Yet nutrigenomic tests are among the first DTC genetic tests and remain among the most controversial. How should firms respond to regulatory scrutiny? One approach is to ask regulators to develop clear guidelines for laboratory analysis, marketing and communications. From an industry standpoint, Gill contends that clear rules, fairly enforced, will prevent irresponsible or unethical uses of nutrigenomics and will increase consumer confidence and trust.

One option is for industry to self-organize and self-regulate. Industry groups, in collaboration with researchers, regulators and consumer organizations as warranted, could establish standards of scientific evidence for commercial offerings of services and products, as well as codes of best practices to ensure test accuracy, fair advertising, client counseling and privacy protection. Regulatory pressure may create an incentive for firms to cooperate and to set standards to govern themselves.

Regulatory pressure arises from gaps and ambiguities in how existing laws apply to a new field like nutrigenomics and the view that those laws are inadequate in scope or enforcement. Hogarth discusses some of these problems and recommends that regulators address differences in regulation of *in vitro* diagnostic kits and laboratory-developed tests, as well as risk classification of genetic tests. Similarly, Ries agrees that categorizing tests is important, particularly if nutrigenomic tests are regulated at the medical end of the spectrum where it is generally assumed that they reveal more serious information, like predictive testing for certain cancers and conditions like Alzheimer's disease. Other tests, like genome-wide scans, reveal less certain information about health risks and are therefore of less medical gravity, but may be viewed as raising more serious consumer protection issues because they arguably represent premature commercialization. Indeed, apart from the regulation of genetic tests, it is a difficult task to determine what kinds of regulations, if any, ought to apply to nutrigenomic health claims. One obstacle is making a decision about the dividing line and balance between providing precise advice that tracks medical advice, and more general and perhaps 'common sense' advice. Saukko points out that until there is a shift away from polarized terms of 'miracle cures' or 'snake oils', the real problem of sorting out what to do with the varying types of genetic tests and advice will not be addressed.

Nutrigenomics may have an impact on the composition of foods, which are themselves regulated in various ways. As Morin points out, food regulators have traditionally had a harm reduction approach that emphasizes protecting consumers from hazards of unsafe food consumption. Newer regulations, however, recognize that some foods can contribute to the maintenance of good health in significant ways and so permit health claims that promote purchase and consumption of specific products. Nutrigenomics now brings the possibility of even more detailed health claims than current generic ones. Yet Morin suggests that food regulators faced with the daunting task of regulating health claims associated with new foods are unlikely to use statutory, regulatory or administrative incentives to encourage manufacturers to obtain and submit nutrigenomic data.

If these different challenges beset the regulator, where does that leave the consumer, especially the growing number of them who seek information, products and services via the Internet? Ries suggests that an online registry of genetic test information is an approach to facilitate greater public access to information about nutrigenomics. A registry would give interested consumers, health professionals and others access to the current state of knowledge about what genetic tests can and cannot say about disease predispositions and the interaction between genetic make-up, nutrient and drug metabolism and other environmental/lifestyle exposures. Companies have an incentive to participate in this type of initiative to show they are not interested simply in short-term financial gain by peddling 'snake oil' to gullible consumers, but are interested in a long-term presence in the market to translate the results of genomics research to consumers and health professionals. Companies that fall into the former category are likely to find that their customer base shrinks as knowledgeable consumers choose to deal with more reputable and transparent companies.

DEVELOPING PROFESSIONAL COMPETENCY IN NUTRIGENOMICS

Nutrigenetic testing is already in the marketplace and it is inevitable that health care professionals will face inquiries from patients or clients seeking guidance on interpreting tests they have purchased. They may also seek advice about whether such tests might be beneficial, especially for genetic tests that are not available through public or private health insurance. Farrell points out that many steps are needed to prepare health care professionals to advise and interpret nutrigenomic tests. Recognizing that medical school curricula to this day do not emphasize genetics, she recommends that undergraduate, medical school and continuing education curricula incorporate a genetics perspective. With respect to continuing education courses to enhance the genetics knowledge of existing practitioners, incentives to attend the courses such as study leaves, in-house workshops, reimbursement for fees and additional certification could be offered. Cassels recommends that nutrigenomics educational initiatives be evaluated to ensure they are comprehensible to health care professionals and provide them with information relevant to their practices. To this end, assessing knowledge levels of health professionals and the effectiveness of communications materials for comprehension, clarity and understanding will be an ongoing requirement. Incentives to join interdisciplinary health care teams could help overcome knowledge gaps in practice; for example, physicians, nurse practitioners and dietitians with specialized training in nutrigenomics may work together to provide a range of skills and expertise to patients or clients.

Among health care professionals for whom nutrigenomics has most relevance, registered dietitians are likely to be early adopters of nutrigenomics into their practice. Undergraduate courses and other training opportunities, including online approaches, would help dietitians gain competence in nutrigenomics. Vogel, DeBusk and Ryan-Harshman, like Farrell, recommend that dietitians work across disciplines and establish strong collaborations with key stakeholders, including government, industries, consumers and the media. At the same time, it may be important for the dietitians' legislated scope of practice to be reconsidered and updated to take into account new services and service delivery models.

COMMUNICATING TO THE PUBLIC

Gill recommends that open dialogue between a coordinated nutrigenomics industry and regulatory representatives will help the industry grow and thrive. She sees this as an opportunity to increase public confidence in nutrigenomics. This raises the question about what types of messages the public currently receives about nutrigenomics. Caulfield et al. observe that popular representations of nutrigenomics are often quite positive, as exemplified by scientific review articles, the popular press and commercial marketing material. They caution that overemphasis of potential benefits of nutrigenomics does not provide an appropriate picture of the state of the science or barriers to achieving claimed benefits. Also, as Gill's point makes clear, it certainly does not correctly portray the interactions between firms and regulators.

Upstream issues in the research, development and regulation of nutrigenomics might inevitably show up in downstream communication about nutrigenomics to the public. Saukko's view is that the firms and regulators must come to terms with a choice in how nutrigenomics is developed and marketed to the public because, at present, firms offering tests oscillate between marketing the tests as 'wellness optimising lifestyle' and 'illness preventing medicine'. In her view, offering general advice on healthy living is innocuous, not particularly harmful and less problematic for firms than if they give clinical-type recommendations. But the former is often criticized as deceptive because it markets generic health advice as personalized by genotype. The other option is to provide advice that is more consistent with traditional medical approaches, an approach that attempts to fulfill the promise of genetics to deliver individualized and specific health advice and treatments. This approach, however, is riskier for firms when the science is at a developmental stage and cannot fully support recommendations for screening or modifying nutrient intake.

If the public receives unclear messages about nutrigenomics, it is because the fundamental choice has not been made about how firms will position themselves and how regulators will respond. Cassels sees an important opportunity and need for further qualitative research that will track the evolution of how nutrigenomics is communicated as it proves its usefulness in health outcomes. He has in mind research that tests the impact of actual print and online media advertising related to nutrigenomics and assesses media material for tone, balance, accuracy and comprehensiveness. Focus group testing of public opinion and understanding of the content and impact of new information targeted at the public will generate feedback about the clarity and coherence of nutrigenomics communications to the public.

MOTIVATING BEHAVIOR CHANGE

The ultimate goal of nutrigenomics is frequently described as the ability to give genetics-based advice to people so they can alter their diet and other lifestyle factors to prevent or mitigate chronic disease. Korthals thinks current views of well-being, health and food are strongly oriented towards health as a personal affair in which the prevention of calculable risks can be achieved by taking personal responsibility and eating well. Well-being is narrowed down to health in this sense. Korthals makes recommendations about whether this is the only sense of well-being that nutrigenomics ought to pursue. Before examining other options for nutrigenomics, one might pause to consider what steps would be needed to fulfill the promise of preventing calculable risks by taking personal responsibility and eating well.

Bouwman and van Woerkum wonder whether giving people genetic information while advising them about how they can modify their diets contributes to better compliance. They believe more research is needed to explore if people actually perceive genetic information – their genes – as relevant to their health with respect to nutrition advice. If people are given dietary recommendations based on personalized nutrigenomics, qualitative research is needed to understand how they would integrate that information into everyday activities of buying, preparing and consuming food. As this research has yet to be conducted, the development and implementation of an evaluation system to monitor changes in eating practices is also a priority. Caulfield et al. also recommend research to demonstrate how consumers will assess and apply nutrigenomics because there are currently few data either for or against the hypothesis that genetic information has any meaningful influence over lifestyle behaviors.

Ries provides a word of caution against thinking that access to direct-to-consumer tests ought to be restricted because there is no evidence that the information motivates behavior change. Many products and services are available for purchase that can provide health benefits if consumers use them in particular ways (for example, fitness equipment, smoking cessation programs), but access is not restricted simply because consumers buy them but do not ultimately make long-term behavior changes. This is fundamentally an issue about deciding which aspects of new products should be left to the marketplace – i.e. relationships between sellers and buyers – and which are the appropriate subject of government regulation.

EXPANDING THE SCOPE OF NUTRIGENOMICS

Nutrigenomics applications presently focus on providing health benefits to proactive individuals. As Caulfield et al. note, population health interventions do not appear to be foremost on the agenda. That is not to say, however, that nutrigenomic science cannot be translated into public health measures, but it is to say that they have not happened yet. Castle identifies three steps that must be taken to put nutrigenomics on a course toward public health. First, the potential of nutrigenomics for public health interventions versus health promotion must be thought through because they lead to different applications of the science. Second, it may be the case that genotyping is a follow-on to the use of already existing public data about disease patterns, allowing public health applications of nutrigenomics targeted at higher risk groups, rather than universal genotyping. Third, the evaluation of the impact of nutrigenomics, mentioned above by Bouwman and van Woerkum, will have to be measured in longitudinal studies to determine if there is public benefit. Decision-makers need to be attuned to the fact that chronic disease reduction will not be settled by public health measures within a term of office, but will require coordinated and sustained effort to make it work, and patience in the collection of evidence.

Korthals challenges many of the assumptions made about nutrigenomics, chief among them the idea that nutrigenomics is necessarily tied to the health promotion of motivated individuals. While his observations lend support to the idea that nutrigenomics can be configured to support the health and well-being of the collective – hence public health applications of nutrigenomics – he also wishes the critique to go further. As he says, there is a risk that nutrigenomics might distort our relations with food such that we think of it only in terms of a narrow conception of health based on risk management. Yet food has social, aesthetic, cultural and other functions that shape individual and group identities, and the way that people interact with food, and with one another with respect to food, gives special meaning to the different functions of food. Korthals advises us to think more broadly, not only about food, but about the potential uses of nutrigenomics to broaden the horizons of these other aspects and functions of food.

Index

S

T

U

V

W

Printed in the United States
By Bookmasters